Security in the Indian Ocean Region

Role of India

Security in the Indian Ocean Region

Role of India

Rockin Th. Singh

Vij Books India Pvt Ltd

NEW DELHI (INDIA)

Published by

Vij Books India Pvt Ltd
(Publishers, Distributors & Importers)
2/19, Ansari Road, Darya Ganj
New Delhi - 110002
Phones: 91-11-65449971, 91-11- 43596460
Fax: 91-11-47340674
web: www.vijbooks.com
e-mail : vijbooks@rediffmail.com

Copyright © 2011, Publisher

ISBN 13 : 978-93-80177-48-9

Price : ₹ 795.00

Printed in India
at Narula Printers, Delhi

Contents

Preface

India sent a strong strategic message across Bay of Bengal with the intention of it is ready to play the role of a security facilitator in the larger Indian Ocean Region (IOR) by generating sonic booms over the Andaman and Nicobar archipelago with Sukhoi-30MKI fighters, and the 3,500-km Agni-III missile creating fireworks off the Orissa coast.

India has legitimate security concerns in the IOR, which falls in its strategic backyard, especially with China making strategic maritime moves in the region. But India might not want to be seen as a regional supercop in the IOR, nor as the prime mover of a naval military bloc in the Asia-Pacific region.

This book also discusses about India considering itself to be "a neutral player" in the entire power-play in the Asia-Pacific region, with the US seeking to "contain" China.

But there is no getting away from the fact that it remains deeply concerned about Beijing's rapid modernisation of its armed forces. The endeavour to build "interoperability" with other friendly Asia-Pacific maritime forces can be gauged from the fact that a "table-top exercise" conducted at the recently concluded multilateral Milan conclave with 12 foreign navies revolved around the security challenges of dealing with the problems of piracy, gunrunning, drug trafficking and illegal migration.

-Editor

1

Geopolitics Of The Indian Ocean

The Indian Ocean and the states on its littoral are of significant and growing importance. The region contains 1/3 of the world's population, 25% of its landmass, 40% of the world's oil and gas reserves. It is the locus of important international sea lines of communication (SLOCs). The region is home to most of the world's Muslim population as well as India, one of the world's likely "rising powers."

The Indian Ocean also is home to the world's two newest nuclear weapons states, India and Pakistan, as well as Iran, which most observers believe has a robust program to acquire nuclear weapons. In addition, the region constitutes one of the key centers of gravity of international terrorism - - "the broad incubator of terrorism" in the words of one conference participant. While India and some a few of the other littoral states appear to be on a path of sustained economic progress, most of the region is characterized by high levels of poverty. The Indian Ocean region suffers from a high level of international and internal conflict and is a key venue for international piracy. It also is the locus of some 70% of the world's natural disasters.

The regional strategic environment is volatile and dangerous. In addition to some of the conditions enumerated above, recent developments in Iraq and Afghanistan now pose additional challenges of violence, terrorism, and instability across the entire Indian Ocean region. A Malaysian conference participant, for example, argued that the foregoing conflicts have been bad for Malaysia and have "played into the hands of local terrorists." For these reasons and others, the region - - an "insecurity community" - - has been an arena of increased diplomatic and military activity on the part of a variety of littoral states as well as external powers in last few years.

Military power, including weapons of mass destruction and their delivery vehicles, is looming larger in the region. Conference participants pointed to the ongoing insertion of military power in the region by the United States and - - to a much lesser extent - - Japan. At the same time, it was emphasized that India, Malaysia and a variety of other littoral states are strengthening their militaries. We also will see "a resurgent Iranian naval capability eventually." Moreover, many of these states are emphasizing power projection capabilities, often through the acquisition of more advanced military hardware and the construction of new bases intended for forward defence.

For example, one Indian participant claimed that India's relatively new Andaman and Nicobar Command - - which New Delhi almost decided to name "Southeast Asia Command" - - is intended to "stop the Chinese east of the Malacca Strait." Quite a few conference participants underscored the growing role of nuclear weapons in the region. The Israeli scholar pointed to Israel's growing emphasis on strategic reach and on the development of a maritime second-strike nuclear capability with respect to both the Indian Ocean - - to deal with Iranian and Pakistani contingencies - - and the Mediterranean.

The Indians, similarly, emphasized their intention to develop a full triad of nuclear weapons capabilities to include, in the words of one scholar, "high yield nuclear weapons in the Indian Ocean." India remains concerned about the role of external powers (or some of them) in the Indian Ocean region. Most of this concern relates to China and - - to a lesser extent, the United States. On the other hand, some conference participants - - to include some Indian participants - - believe that key littoral countries like India, Pakistan, Iran, and Malaysia have gained substantial space and strategic autonomy because of the desire of external powers to forge alliances and coalitions in the region. One Indian, moreover, commented: "asking outside powers to stay away is a pipedream."

The Indians present also seemed comfortable with, and appreciative of, Israel's expanding security perimeter and its growing strategic involvement in the Indian Ocean region. The Indians present welcomed the evolving Indo-Israeli security nexus and were pleased by a presentation by an Israeli scholar, which placed India and Israel in a common democratic and civilizational community. Similarly, one Indian - - referring to Moscow's recent naval foray into the Indian Ocean - - stated: "India is pleased that Russia is back in the Indian Ocean."

The region is characterized by growing strategic competition involving both external powers and the littoral states. In this regard, most conference participants emphasized the continuing rivalry between India and China, the "peer powers of Asia," and the potential for this problem to worsen. The Indians at this conference were especially vocal and alarmed about Beijing's evolving role the Indian Ocean region. One Indian, for example, asserted that the 21st century would be the "template for Sino-Indian rivalry."

Pointing to Chinese proliferation of WMD, provision of conventional arms to various South Asian states, "ruthless subordination of its neighbors", "special relationships" with Pakistan and Burma, "growing presence of the PLA" in areas adjacent to India's borders, and developing naval capabilities, most of the Indians present made it clear that China, in their view, is India's number one security problem. Commenting on India's insertion of naval forces into the South China Sea, one Indian said it was a "good thing if China felt threatened by our exercise. We intended to send a message and they got the message."

The Chinese scholar present strongly countered these assertions and argued that Chinese strategy in the Indian Ocean is benign and has three dimensions: trade and development, good neighborliness and friendship, and security and cooperation. In his view, China has re-oriented its overall security strategy since the end of the Cold War, but India has not done so. He argued that China is no longer preoccupied by fears that other states are "encircling" China, but Indian national security strategists remain fixated on fears of encirclement of India.

Some of the Americans at the conference also attempted to calm Indian fears, arguing, for example, that China's security strategy is oriented mainly east, not south. The Indians generally reacted to such interventions with skepticism. In addition, on the matter of Burma, one conference participant argued that fears about Chinese influence in Burma are overdrawn and it is not Beijing, but Rangoon, that holds the "whip hand" in the China-Burma relationship.

Paralleling this concern with China, there also is some Indian worry about the growing role of the United States and, to a lesser extent, Japan in the region. At the extreme, one Indian argued strongly for the need for Indian military - - and nuclear - - contingency plans with respect to a potential U.S. threat. This worry about the United States and U.S. power is so notwithstanding the reality that almost all of the Indians at the conference welcomed the development of closer Indian ties with the United States.

According to an Indian naval officer, "the United States is unquestionably the most dominant player in the Indian Ocean in modern history." The United States has the capability to project military power in the region and a well-defined strategy to pursue its policy of preeminence. The U.S. maritime strategy of the 1980s envisioned a war at sea won by sea control.

The new U.S. strategic thrust aims to move away from classical sea control/sea denial to influencing events further ashore as exemplified by Afghanistan. He observed: "if the U.S. objective is to share with ...other nations ...(the) enduring objective...(of) stopping the emergence of a hostile coalition, then it must remain engaged in the region. Its policies must work towards engagement rather than adopt unilateralist approaches." The Indians present also seemed anxious about how the United States will "manage" China when the PLA navy or other military forces start to operate in the Indian Ocean, as New Delhi believes likely.

India will be increasingly attentive to its interests in the Indian Ocean region in the coming years. This is suggested by a variety of considerations. All Indian participants in the conference stressed the importance of the Indian Ocean to India from economic, political, legal and military perspectives. India's political and naval leadership is convinced that matters maritime are going to play an increasingly important and critical role. India needs a secure maritime environment to achieve sustained national development.

In addition, many Indians see the Indian Ocean as India's backyard and see it as both natural and desirable for India to function as the leader and the predominant influence in this region - - the world's only region and ocean named after a single state. To this end, there was broad agreement among the Indian delegation that India's security perimeter - - its "rightful domain" - - extends from the Strait of Malacca to the Strait of Hormuz and from the coast of Africa to the western shores of Australia. India, according to a senior Indian naval officer at the conference, "will have to play a very large role (in the Indian Ocean) if the prospects for peace and cooperation are to grow."

India will try to exert a strong hand in this region for fundamental national security reasons. Protecting India's EEZ of over 2.3 million square kilometers, securing India's energy lifelines, promoting overseas markets and fulfilling international commitments are some of the interests to which India is sensitive. As expressed at the conference, New Delhi's "Look East" policy, its growing ties with Israel and Africa, and even Iran, and its naval, air and nuclear weapons modernization efforts, all are related to these concerns. Aside from India, many of the other littoral states are acquiring a more pronounced maritime orientation and developing closer links with one another.

Malaysia, for example, is more focused now than ever before on the potential strategic importance of the Indian Ocean approaches to Peninsula Malaysia. Not long ago, Malaysia's navy chief said that the country's strategic location in the waterways of the South China Sea and the Indian Ocean exposes the country to serious dangers. Reacting to this challenge, the Malaysian navy has inaugurated construction of a new navy base and command center at Langkawi, Kuala Lumpur's only port directly fronting the Indian Ocean.

Thailand, similarly, is now more aware of its status as an Indian Ocean littoral state. Arms trafficking in southern Thailand, which has fueled conflicts in Sri Lanka and northeast India, has come under scrutiny as Thailand's neighbors have urged a more robust response from Bangkok. In recent years, Bangkok also has joined a plethora of Indian Ocean regional organizations - - including BIMSTEC and IOR-ARC, and has pursued the so-called "Look West" policy of cultivating Indian Ocean states, especially India.

Thailand lately has also shown new interest in building a canal across the Kra Isthmus to forge a shorter direct route between the Pacific and Indian Oceans. However, large obstacles stand in the way of this dream being realized any time soon, not the least of which is Singapore's implacable opposition to a Kra Canal.

Sub-regional efforts to promote Indian Ocean peace and security are more likely to bear fruit than are region-wide schemes. To this end, conference participants were of one mind that confidence-building and similar measures would be most successful if attempted in Bay of Bengal (the area of operations of BIMSTEC) or the Arabian Sea or between the Indian and Pakistani navies. On the other hand, large region-wide efforts such as the "Indian Ocean Zone of Peace" concept or even the Indian Ocean Region Association for Regional Cooperation are much less likely to succeed.

THE CONCEPT OF AN INDIAN OCEAN REGION

The academic study of the international relations of regions is largely a post—World War II development. It may be traced in part to the emergence in the 1940s and 1950s of the "area studies" approach, inspired more by a particular interest in the affairs of a given locality than by a general interest in global affairs. Western Europe achieved a modern regional identity during postwar rehabilitative efforts. The coming to independence in the fifties and sixties of large numbers of former colonies generated additional regional consciousness, with Southeast Asia and parts of Africa being added to Latin America and the Middle East as areas of study by regional experts.

Furthermore, the "loosening of the bipolar world, moves toward autonomous policies by the middle range powers, and explicit efforts at fostering patterns of international collaboration in various areas of the world" all served to direct scholarly attention toward "regional foci of interaction." Regions have also come to be seen as useful intermediate units of analysis at a level between individual nation-states and the world as a whole. There is a modest body of literature on the international politics of regions that draws upon concepts of general systems theory and the research tradition of systems analysis.

Working from the conception of the world as "the international system," a group of scholars has applied the systems perspective to analyses of geographically distinct (and otherwise distinct) groupings of states, variously called "subordinate systems," "subsystems," or "regional subsystems" of the international or global system.

Generally recognized as the first to attempt such an approach was Leonard Binder in a 1958 article on the Middle East. Following from Binder's pioneering effort have been a number of works seeking to submit area data to systematic analysis. Characteristically, these subsystem studies have considered both disintegrative and integrative developments—both conflict and cooperation. They have gone beyond the area studies tradition of intense interest in one particular area for its own sake toward a "heightened interest in the relationships between the global system and regional subsystems," relationships virtually unexplored by area specialists.

Writing in 1973, William R. Thompson surveyed the extant international relations literature that had used the sub-systemic approach to the study of regions. He concluded his article with an explication of the concept of "regional subsystem" based on his analysis of twenty-two academic works in which the concept had been centrally applied. Thompson concluded that there were four necessary and sufficient conditions for identifying a regional subsystem: "The actors' pattern of relations or interactions exhibit a particular degree of regularity and intensity to the extent that a change at one point in the subsystem affects other points.

The actors are generally proximate. Internal and external observers and actors recognize the subsystem as a distinctive area or 'theatre of operation.' The subsystem logically consists of at least two and quite probably more actors." It will be argued in this chapter that there is in the Indian Ocean a discernible "linkage of instabilities" such that "a change at one point ... affects other points"—the first condition above. Furthermore, evidence will be cited for increasing levels of economic interaction and cooperation among Indian Ocean states, as well as for incipient security cooperation among various groups of states.

With respect to Thompson's third condition, it seems clear from both internal and external perspectives that the Indian Ocean is now recognized as a distinctive "theatre of operation." The superpowers, in particular, have increasingly approached the Indian Ocean area as a "strategic arena" for the multiplicity of reasons discussed in the second section of this chapter. For their part, the Indian Ocean states have become increasingly sensitive to external intrusions, both as threats and as opportunities. The Indian Ocean zone-of-peace and nuclear-free-zone initiatives and the existence and longevity of the United Nations Ad Hoc Committee on the Indian Ocean constitute some of the evidence of an indigenous perception that the Indian Ocean defines a distinctive area in international politics.

Thompson's fourth criterion of two or more actors is prima facie no obstacle to considering the Indian Ocean area a regional subsystem. Indeed, some critics may object that thirty-six actors are too many, but Thompson set no upper limit and other widely accepted regional subsystems have even more "members." Whether the Indian Ocean area meets Thompson's second definitional criterion for the existence of a regional subsystem is more problematic. Whether South Africa, Iran, and Australia can be said to be "generally proximate" is highly questionable by most standards. Indeed, most critics of the notion that the Indian Ocean defines a "region" argue that the area is simply too big for such a conception to be meaningful.

A solution to this dilemma of size and expanse is provided by Cantori and Spiegel's concept of "core sectors" within regional subsystems. A core sector "consists of a ... group of states which form a central focus of the international politics within a given region.... There may be more than one core sector within a given subordinate system." It is possible to identify at least five core sectors within the Indian Ocean regional subsystem: a Persian Gulf core, a South Asia core, a Red Sea core, a southern Africa core, and an Australasia core. Some state actors may usefully be considered as members of more than one core.

Saudi Arabia, for example, is clearly a principal Persian Gulf actor but it has also from time to time been heavily involved in the Red Sea core, defined as the two Yemens, the Horn of Africa (Ethiopia, Somalia, Djibouti), and, on some issues, Sudan and Egypt. To take another example of multiple core membership, Pakistan—a "natural" member of the South Asia core—has become increasingly involved in affairs in the Persian Gulf, in the spirit of Islamic solidarity and for the reality of balance-of-payments benefits. An East Africa core was apparent in the late 1960s during the most successful period in the life of the East African Community. Kenya recently has been drawn more into the affairs of the Horn of Africa; Tanzania, a "Frontline State," into the affairs of southern Africa.

It is not inconceivable, however, that an East Africa core will some day reemerge when political and economic factors are once again conducive to higher levels of interaction in that sector of the Indian Ocean littoral. So what? What is the rationale for such a conception of an Indian Ocean regional subsystem with multiple core sectors—that is, what purpose does it serve? It is potentially useful in at least two ways: first, to assist the analyst (and maybe even the policy maker) in thinking about international politics and security issues in that part of the world and, second, to aid in the description of reality.

Of course, those two potential benefits of a regional systems perspective on Indian Ocean affairs are not mutually exclusive. To the contrary, thinking about the world and describing the world (or portions and aspects of it) are really sequential occupations on a continuum familiar to all those who aspire in some sense to be "scientists." The next two steps are explanation and prediction. With regard to the first potential benefit of a region/cores conceptualization, namely, its use in thinking about the politics and security of the Indian Ocean area, the systems-analytical approach advocated here is neither magical nor mysterious (despite the efforts of some of its jargonistic exponents to shroud it in mystery—or obfuscation).

Systems analysis, according to Michael Banks, is simply "a more formalized version of clear thinking about complicated problems.... We divide a large problem into sections, concentrate our attention separately and singly on each section in turn or on a group of sections, and we explain each part to ourselves, rebuild[ing] the whole piece-by-piece in order to reconstruct the phenomenon mentally in a form in which we feel we can understand it."

As Richard Little points out, "There is no body of rules indicating how a systems approach should be implemented. There is, therefore, no formal methodological procedure associated with the approach. Nevertheless, there is a systems perspective and it is normally quite clear when analysis is being written from this perspective."

Characteristic of the systems perspective advocated in this chapter are the following assumptions:

1. A system (or regional subsystem) is more than simply the sum of its parts. It is both a "structure" (construct) consisting of its components (e.g., the Indian Ocean nation-states) *and* the transactions among and between those parts. Much of the essence of international politics consists of linkages, interactions, reactions, and interdependence—more than the simple sum of all relevant national foreign policies.

2. Various actors (individual decision makers, nation-states, multinational corporations, international organizations, etc.) are assumed to be conditioned and constrained by the characteristics of the system in which they operate. In other words, "systems-level forces seem to be at work." Therefore some part of the explanation of international behavior and

of policy outcomes is to be found in the characteristics of the system. Furthermore, influences and constraints are assumed to flow in both directions: just as the structure of the system affects interacting units, so too do the actions of the units affect the system's structure. The interrelationship is dynamic and reciprocal.

3. The examination of patterns of international relations on various levels (e.g., subnational, national, regional levels) contributes to an understanding of politics at the global level.

4. Even in the field of international relations, there are areas of coherence and orderliness in the midst of apparent randomness and diversity. "[A system] is a means of organizing apparently chaotic behavior between entities." Systems thinking are meant to be "an attack on the problem of complexity."

5. Finally, systems thinking can be used as a bridge to insights from other social science disciplines, such as political geography. "Political geographers have been accustomed to thinking in terms of system relationships almost from the beginning of their field. Thus a system framework ... will be easily understood by the political geographer, and his work easily adapted to it."

The second potential usefulness of the regional systems perspective, namely, to abet description (and to contribute toward explanation) of the reality of international relations in the Indian Ocean region, arises from the suggestiveness of systems thinking. It inspires propositions that can be tested against reality. For example, the definitional assumption that systems are dynamic (not static) entities suggests questions about the nature of system transformation. Is the system becoming more cohesive or is it disintegrating? Are its interactions becoming more cooperative or more conflictual? Operationalizing degrees of cohesiveness and levels of cooperation is not easy but it is possible.

More controversial has been research into the issue of stability within a system, particularly as it relates to the structure and distribution of power. For example, an interesting question with respect to the Indian Ocean subsystem, or its various cores ("sub-subsystems"), is whether stability is increased by the concentration and hierarchical distribution of power or by the diffusion and roughly equal distribution of power.

The first proposition is intuitively more appealing. The Persian Gulf core was a more stable place when the shah's Iran dominated (or was perceived by other core actors to dominate) the power hierarchy. In South Asia, the clear predominance of India may have contributed to a more stable core area than might otherwise have been the case (particularly since Indian predominance became so clearly apparent after the birth of Bangladesh). The southern Africa core is another case in point with a dominant South Africa.

The work of David Singer and various associates provides potentially useful operational techniques for research into the question of power distribution versus stability in the Indian Ocean region. The multilevel characteristics of the systems perspective suggest and abet inquiry into additional issues such as: the opportunities for subsystem dominance (e.g., Persian Gulf oil producers versus "dominant system" consumers); Islamic revivalism (e.g., in Iran, in the Persian Gulf core, in the Indian Ocean generally); and the politics of ethnicity (e.g., Kurdish insurgency in subnational areas of Iraq and Iran, liberation movements in the southern Africa core).

Finally, systemwide problems suggest the need to search for systemic solutions. Such problems as the preservation and management of migratory fish stocks, the necessity for joint action against sources of undesirable levels of air or ocean pollution, and the curtailment of militarization or nuclearization are likely to lead to (and, indeed, have led to) a convergence between the concept and the reality of an Indian Ocean regional system. It is to an examination of such an emerging reality that we now direct our attention.

REAL WORLD BASES

One reason why scholars—particularly those specializing in international relations, strategic studies, and foreign policy—have increasingly begun to think in terms of an Indian Ocean region is because both external and indigenous policy makers have themselves been approaching issues in that area in a more comprehensive fashion.

EXTERNAL PERSPECTIVES

Current superpower rivalry in the Indian Ocean arena is, in a sense, business as usual. Only the players have changed, and the stakes. The "post-Gaman" history of that part of the world has witnessed successive hegemonies by external (European) powers. Since 1498, the Indian Ocean has had a sort of strategic unity or coherence imposed from the outside by, successively, Portugal, Holland, and Britain, with France, Germany, and Italy challenging British predominance in certain sectors at certain times with limited success.

The Portuguese empire in the Indian Ocean was established in accordance with the strategic plan of Alfonso d'Albuquerque. His strategy included capturing the approaches to the ocean, sealing off entrances to foreign shipping, and establishing bases along the littoral.

Albuquerque secured the cape route by occupying key points along the East African coast and he controlled the entrance to the Red Sea by capturing Socotra Island. Hormuz was taken in order to dominate the Persian Gulf, and Malacca to command access to the Spice Islands (Malaysia and Indonesia). The same "choke points" were objectives of subsequent imperial contestants, with the Suez

Canal added to the list in 1869. The contemporary "base race" of the superpowers, seeking friends and real estate near the same choke points, is not unfamiliar against this historical background.

Holland ended Portugal's predominance in 1641 by capturing Malacca. In 1652, the Dutch established the first European settlement on the Cape of Good Hope, a possibility that, curiously, the Portuguese had overlooked. According to Auguste Toussaint, "The period from the fall of Malacca [to the Dutch in 1641] to the completion of the British conquests in 1815 was really one long interregnum during which no single power controlled the ocean."

During that period, the principal struggle for control was between the British and the French. But for over 150 years after the Congress of Vienna, the Indian Ocean was essentially a "British lake."

While other European powers obtained or maintained footholds at various locations in the region, Britain held all the most strategic points and tied them together with the Royal Navy. The explosion of nationalist sentiment in the wake of World War II precipitated a rapid and sometimes cataclysmic decolonization process in the area. Beginning with the partition of India in 1947, it had largely run its course by the early 1960s. The British announcement in 1968 of withdrawal from "east of Suez" by the end of 1971 marked the end of the era of British hegemony in the Indian Ocean.

The superpower competition in the area that has escalated since the British withdrawal can be seen in historical perspective as yet another attempt by external powers to establish a strategic condominium over the Indian Ocean, albeit for different reasons and in a substantially transformed environment. Washington and Moscow, and a few other major capitals, have developed de facto Indian Ocean policies, though not always explicit ones. Indeed, if the case for the contemporary strategic coherence of the Indian Ocean area depended on the existence of Indian

Ocean desks and sections in foreign ministries and defence bureaucracies, there would be little if anything more to say.

That is, in the governments of external powers the responsibility for Indian Ocean policy is shared, usually among policy makers working on Africa, the Middle East, South Asia, and Oceania, with no single office having overall authority for policy development toward the region per se. Of course there is nothing particularly unusual about this situation.

Policy making is invariably a fragmented process in which the burden of formulation is distributed among and within different departments of government. This reality of bureaucratic life therefore suggests that we direct our attention away from policy formulation to policy outputs and their focus. What policies have external powers actually pursued toward the region?

The policies of the two superpowers are naturally of primary concern. Their immediate interests in the area have merged with the broader conduct of their global rivalry so that the Indian Ocean has become a strategic arena of considerable importance to both. The interests of the United States revolve around the need to ensure access to Persian Gulf oil for itself and its allies. American dependence on gulf oil has never been as great as that of Western Europe and Japan (which received approximately 75 percent and 90 percent of their oil, respectively, from the gulf at the time of the 1973-74 embargo).

The most portentous result of the embargo was not the damage to Western economies (which was considerable), but the serious bickering and back stabbing in the Atlantic alliance as member states scrambled for favored access to unembargoed oil and for future access to Arab oil. A sustained denial of Persian Gulf oil to the West is an eventuality that the United States is therefore keen to avoid for reasons of alliance solidarity as well as economic health, a fact underlined by the Carter Doctrine and the development of the Rapid Deployment Force (RDF).

Despite the recent glut in worldwide oil supplies and falling prices, Persian Gulf oil—about 55 percent of the world's proven reserves—will remain of vital strategic importance to the West through the end of this century. A distinct but related interest is the concern of the United States to maintain sea lines of communication (SLOC), particularly through vital choke points such as the Straits of Malacca, Babel Mandeb, and Hormuz. For years the United States, consistent with the Nixon Doctrine, depended upon local surrogates to defend its core political, economic, and strategic interests in the Indian Ocean area.

But in the wake of the events that began to shake the region in the late 1970s Washington developed a new approach to its security interests there. By actively seeking to enhance its access to naval, air, and communications facilities throughout the area, by improving the operational capability of British owned Diego Garcia, by increasing the level of Indian Ocean naval deployments, and by creating the RDF, the United States has substantially improved its capacity to project power into the Indian Ocean and has thereby declared its intention to take a more active role in the region's affairs.

Whether these military responses are appropriate to what many see as essentially political and social problems in the area is a question of profound importance but it is beyond the scope of this chapter. The question at hand is whether the actions of American policy makers evince a coherent strategic approach to the region. Members of the Reagan administration certainly have argued that they do present a unified approach to policy. In 1981, for example, the secretary of state noted that "our broad strategic view of the Middle East recognizes the intimate connections between the region and adjacent areas: Afghanistan and South Asia, northern Africa and the Horn, and the Mediterranean and the Indian Ocean."

Similarly the deputy assistant secretary of state for Near Eastern and South Asian Affairs testified that "our approach takes into account threats and developments in contiguous areas. We will carry out a coherent and consistent policy in full awareness of the interrelationships between tensions in different regions and theaters." The claims of any government to be acting consistently on the basis of a coherent policy should, of course, be treated with utmost caution. Nevertheless, there does appear to be a relatively high degree of coherence, at least in American declaratory policy, toward the Indian Ocean.

The efforts of the United States to improve its position in the area are focused, not surprisingly, on the Persian Gulf, but it would be a mistake to conclude that the policy is exclusively gulf-centered. The concentration of efforts in the northwest quadrant of the Indian Ocean reflects the high priority that Washington attaches to its interests in the gulf and an overall emphasis on the stability of that core sector. However, Washington has not overlooked its political and military concerns in other parts of the region. Thus, in the last few years it has sought to improve its military dispositions in the eastern approaches to the Indian Ocean by, among other things, securing landing rights for B-52 bombers in Australia.

In the political sphere, area wide policy concerns have been reflected in assiduous attempts to secure better relations with India and in closer contacts with South Africa. Despite some inconsistencies, American policy makers convey both by their words and their deeds an overall perception of the interdependence of events in the Indian Ocean region. There is a recognizable framework for American policy in the area that, while arguably inappropriate to the political forces at work in the region, nevertheless suggests a coherent and comprehensive approach to the protection of American interests.

Finally, it is noteworthy that the Russia appears to recognize a certain coherence in American policy. In April 1979, *Pravda* referred to "the defence line being created by the Pentagon along the Egypt-Israeli, Persian Gulf, Diego

Garcia, Australian perimeter." In contrast, the policies of the USSR itself exhibit a more inconsistent quality. Much of this appears to be attributable to its general inability to gain support for its policies rather than to an absence of any clearly defined interests in the region. In fact, the Soviet Union's proximity to the region has produced a certain continuity of interests that continues to dictate the course of Soviet policy as it has in the past.

Foremost among those interests is Moscow's preoccupation with the maintenance of stability on its borders and of a measure of influence, if not control, over its neighbors. This standard dimension of Soviet policy is most clearly manifested in its relations with the states of Eastern Europe but it also has relevance throughout Soviet Asia where ancient cultural and ethnic traditions tend to undermine Moscow's political authority and create natural communities of interest with peoples outside the Soviet Union. The Iranian revolution and the Iraq-Iran war are particularly worrisome to Moscow because they could give rise to unstable or anti-Soviet regimes on its borders. At least the shah was predictable.

Apart from the proximate territory to the south, the Indian Ocean itself is of importance to Soviet security. It offers one alternative means of linking Soviet Europe with Soviet Asia should the trans-Siberian railroad be rendered inoperable in peace or war. The Indian Ocean also offers a back door to China by which the Russia could relieve military pressure along their common central Asian border if necessary. Thus, like the United States, the Russia has an abiding interest in maintaining sea lines of communication throughout the region. Moscow also sees the ocean as a potential operational area for American strategic missile submarines that should not be allowed to move about unchallenged.

Finally, the Indian Ocean is an arena in which the Russia competes for influence with the United States as part of the global search for strategic advantage. It has discovered that in addition to hunting submarines, warships can be used to reassure friends and to discourage potential enemies. This catalogue of interests continues to provide a foundation for Soviet policies in the Indian Ocean region. From the Chinese perspective at least, these policies have been seen as exhibiting a regional coherence, a view that lends support to the notion of a Soviet Indian Ocean policy. According to the Chinese, "Moscow is stepping up its strategic dispositions along the arc from Africa through West Asia to Southeast Asia."

Beijing is hardly an objective observer; even so, there appears to be substance to the Chinese analysis. Over the past decade the Russia has attempted to expand its influence throughout Africa, the Middle East, and Asia, while maintaining a relatively high level of naval deployment in the Indian Ocean. In the early 1970s its objective was a collective security regime that would embrace much of the region. This failed, however, to attract the support of local states and is recognized, even by Soviet analysts, as unattainable at present. Moscow also has suffered setbacks in its bilateral relations in the region, having been ejected from Somalia and Egypt and "forced" to intervene in Afghanistan.

Overall, the pattern of Soviet activity in the region is fragmented but care should be taken in suggesting that Soviet policies are similarly fragmented. The internally inconsistent features of Soviet policies are probably more a reflection of their mixed success than the consequence of a basic lack of coherence in overall design. The Russia is characteristically opportunistic in the conduct of its foreign policy. This is always likely to create uncertainty among observers about policy objectives. But this opportunism is arguably more closely related to the tactical than to the strategic side of Soviet security policy.

The latter gains its coherence from the constancy of Soviet interests in the region. Besides the superpowers, France, Britain, Japan, and China have substantial interests in the Indian Ocean area. The diversity and relative importance of these interests is reflected in the varying degrees of coherence that is apparent in the policies of these nations toward the region. Of the four, French policy perhaps exhibits the greatest cogency. The French remain quite active in the Indian Ocean, maintaining the largest naval presence after that of the United States and the Soviet Union. Réunion, near Mauritius, and the island of Mayotte (Mahoré) in the Comoros are administratively parts of France, and Paris continues to station army and air forces in Djibouti.

Britain's former status as the principal colonial power has left it with a range of political, economic, and strategic concerns in the region that remain of considerable importance despite its formal withdrawal from east of Suez. Britain lacks the resources to protect independently all of its interests in the Indian Ocean and therefore has to rely heavily on the United States for this purpose.

Nevertheless, the British government, drawing on a rich bureaucratic memory, appears to retain a clear conception of the complicated relationships that characterize the region's affairs. While the evidence may be inconclusive, it does seem that the external powers with the greatest interests in the Indian Ocean region look upon events in one core area as having implications for others. On balance, their policies toward the region reflect a perception of the ocean's strategic integrity.

INTERNAL PERSPECTIVES

If the last decade or so has brought a heightened awareness of Indian Ocean challenges and interests to the capitals of major external powers, particularly the superpowers, it also has resulted in greater apparent understanding of their common dangers and opportunities on the part of indigenous leaderships in the region itself. This is reinforced by a sense of shared identity, based in part on the common historical experience of European imperialism. However, long before the age of imperialism the ocean itself had become a medium of contact, of movement, of exchange, bringing together peoples and cultures that otherwise would have remained isolated from each other.

As A. P. S. Bindra writes, "Milleniums before Columbus traversed the Atlantic ... and before Magellan circled the globe, the Indian Ocean had become a ... cultural highway." The present residents of Madagascar are believed to have originated principally in the Indonesian islands, having arrived in the western ocean in a series of migratory waves. There are large populations of people in eastern and southern Africa whose ethnic roots can be traced back to the Indian subcontinent and the Malay Peninsula.

Arabs and other Muslims historically established themselves along the Red Sea and East African coasts and eventually in the Indian subcontinent and Indonesian archipelago as well. The Hindus themselves had established a sort of "greater India" to the east before being supplanted by the Muslims.

The islands in the middle of the Indian Ocean also have served as meeting grounds. Blacks from the African coast have mixed with south Asians to produce the Creole populations of Mauritius, the Seychelles, etc. The people of the Comoros speak a Bantu-like language with Arab borrowings. The Swahili dialect of East Africa is said to have clear affinities with the Arabic of the Persian Gulf.

In short, there has clearly been extensive contact among the littoral and island peoples of the Indian Ocean. Whole populations have cultural memories and cultural reminders of other Indian Ocean lands. Assimilated island populations are composed more of "Indian Ocean people" than of blacks or Indians or Arabs. Against this historical and cultural backdrop, contemporary strategic and economic relationships are emerging. The proliferation of regional strategic linkages underlines the increasing interrelatedness of events in the Indian Ocean area. Ian

Clark has pointed out that events may be connected in two ways. The first is by means of a "linkage of instabilities," a notion posited by Ferenc Váli in one of the few single author books on the Indian Ocean.

According to Váli, the "instabilities and unbalanced situations which prevail in one subregion not only radiate into the neighboring countries but may also reach out into more distant parts of the region [to form] a 'linkage of instabilities' which extends throughout the area." Situations may "radiate" and "reach out" in a variety of ways, not necessarily uniformly.

The impact of a situation on events near its place of occurrence is likely to be different from that which is felt further afield but, while different, the implications may be highly significant for all parties. Thus, the Vietnamese invasion of Kampuchea created a massive refugee problem and an apprehension of direct threat to security in Thailand, while in places like Indonesia and Australia the invasion was seen as an unwelcome and dangerous manifestation of regional instability.

Similarly, the Soviet invasion of Afghanistan raised the specter of a direct Soviet threat to Pakistan and to Persian Gulf oil fields while raising fears in the more distant parts of the region over the means and ends of Soviet policy. Ill treatment by East African regimes of their citizens of South Asian extraction has periodically caused considerable consternation in the Indian subcontinent.

As a consequence of these and other developments over the past decade, it is now unlikely that any event threatening the interests and security cf even the smallest states in the Indian Ocean will be dismissed as of no consequence to other states of the area. In the second place, Clark suggests that "linkages are created between [sub]regions when individual states, or groups of states, consciously pursue security policies on a wider than subregional basis."

One example is the concerted effort of the black states of southern Africa, together with other Indian Ocean countries, to isolate and place pressure on white-ruled South Africa. A second example is Saudi Arabia's policy of seeking "to integrate security postures from the Red Sea through the Gulf and to the extremities of the Indian subcontinent." However, the linkage process is a fragmentary one at present. The states of one core of the Indian Ocean region have not generally sought security by concerted actions with those of other cores.

Security linkages heretofore have been largely extraregional, involving a local state and an outside power: Australia with the United States, India with the Soviet Union, Pakistan with China and the United States. But there are precedents for states of one core joining a formal alliance with those of another. Although both have included outside powers, the South-East Asia Treaty Organization (SEATO) and the Central Treaty Organization (CENTO) were alliances of this character. Their success, however limited, suggests that more formal core coalitions may be possible in the future.

The recently established Gulf Cooperation Council (GCC), ostensibly a cooperative economic initiative, is evolving in the direction of a security alliance in the context of the Iraq-Iran war, and there are indications that Islamic Pakistan may someday formalize its present de facto involvement in the security of the Arabian Peninsula. Another criterion that can be used to assess internal perceptions of the strategic coherence of the Indian Ocean region is the character of regional responses to external intrusions. Have local states *as a group* adopted common policies and positions in an effort either to deny or to accommodate external (mainly superpower) activities in the region? Among the issues that offer some insight into this matter is the proposal that the Indian Ocean be declared a zone of peace.

This idea was originally advocated by Sri Lanka during the twenty-sixth session of the United Nations General Assembly in 1971. The UN adopted a resolution supporting the proposal and called upon the interested states to enter into consultations to implement it. Since then, all efforts to resolve the problems confronting the proposal have proved fruitless and, accordingly, it remains no closer to reality than it was in 1971. The significant point for this analysis, however, is the initiative itself which demonstrated a collective identity and a collective concern among the Indian Ocean states.

The vision behind the proposal was provided by Prime Minister Bandaranaike of Sri Lanka, who told a Commonwealth heads-of-government meeting in Singapore in January 1971: "The Indian Ocean is a region of low solidarities or community of interests. Although it forms a geographical and historical entity, there are few cooperative links between countries in the region, and these are either bilateral or sub-regional. A Peace Zone in the Indian Ocean will provide countries of this region with time to develop trends toward integration and cooperation so that in course of time the Indian Ocean region could move from an area of low solidarity to an area of high solidarity."

In 1972, the UN General Assembly appointed an Ad Hoc Committee on the Indian Ocean to consider ways of implementing the zone of-peace resolution. The committee has become the principal focus for work on the issue, expanding its membership to forty-eight by 1985. Its efforts have been paralleled by informal meetings of the representatives of Indian Ocean states, initiated by Sri Lanka in 1973 and culminating in a general meeting of littoral and hinterland states in July 1979. Another is in prospect in 1986.

Whatever the ultimate fate of the zone-of-peace concept and a companion proposal for an Indian Ocean nuclear-weapons-free zone put forward by Pakistan in 1974, such initiatives have contributed to a perception of the Indian Ocean region as a distinctive geostrategic zone. Indeed, such a perception can even be said to have been institutionalized in the UN Ad Hoc Committee, which continues to meet annually. Although the Indian Ocean area is one of considerable economic diversity, various aspects of its economic life also lend a measure of support to the notion of emerging regionality.

Here three of these characteristics are examined: the similar economic profiles of most of the states of the area, the movements toward subregional economic cooperation, and the trend toward expansion of intraregional trade. Over half of the states of the Third World are located on the littoral or in the hinterland of the Indian Ocean. Recent World Bank statistics describe thirty of the thirty-six Indian Ocean states as less developed countries (LDCS) with per capita gross national products (GNPS) of less than $4,830. Only one of the remaining nations, Australia, is regarded as industrialized by the World Bank.

The others—Iraq, Kuwait, Qatar, Saudi Arabia, and the United Arab Emirates—are categorized as "capital-surplus oil exporters" (CSOES). This economic profile has had a profound effect on the economic life of the region. Indeed, it can be argued that the comparative economic homogeneity of the Indian Ocean states gives the whole region a degree of coherence as an area of underdevelopment.

At first glance this homogeneity of the LDCS of the Indian Ocean is not readily apparent. There is considerable diversity and disparity among such economic indicators as level of income, rate of economic growth, and size of gross national product.

Bangladesh, the poorest country of the region, has a per capita GNP of only $90, while that of Singapore, a near neighbor, is $3,830. Similarly, India's gross domestic product (GDP) of $112 billion is many times higher than Somalia's at $1 billion. Such statistics, however, tend to obscure some underlying similarities in the structures of the region's economies.

About two-thirds of the LDCS and CSOES of the region have either one-crop or two-crop economies. That is to say, in excess of half—and in most cases much more—of their income is derived from the export of one or two commodities.

Of the LDCS that remain, most depend on only three or four major exports. The absence of diversity in the economies of the LDCS and CSOES means that they have limited flexibility and are subject to similar stresses and strains. For most LDCS, the agricultural sector of the economy makes the largest contribution to GDP. In some countries, such as Somalia, this can be as high as 60 percent but for most the figure is around 30 percent. The service, industrial and manufacturing sectors follow in roughly that order. The CSOES, with their dependence on crude oil production, are in an analogous situation.

In their case, the level of dependence on one sector of the economy is around 75 percent and, in some cases, even higher. In the picture that emerges of the Indian Ocean region, most of the countries have narrowly based economies, highly vulnerable to the vicissitudes of the international economic system. Whether the single "crop" is oil, copper, coffee, sugar, cotton, or some other product, these countries share common concerns regarding access to reliable markets, maintenance of high prices, and sustaining demand for their products. The problems they confront in attempting to achieve these goals could hardly be more evident than at present.

In a period of long-term global recession, both the single-crop agricultural economies of countries such as Mauritius and Somalia and those of the oil-exporting countries have been similarly afflicted with economic decline. In the latter case, these countries now confront conditions thought to have been left behind with the phenomenal oil price increases of 1973 and beyond.

These economic realities have significant policy implications, frequently affecting the positions that LDCS have taken on a range of international economic matters. Thus, on issues of international finance, the transfer of technology, trade liberalization, and, of course, the creation of a "new international economic order," they have similar attitudes and have adopted similar negotiating positions.

In sum, the LDCS and, to a lesser extent, the CSOES of the Indian Ocean region share a set of affinities that are partly obscured by conventional economic indicators. Their governments face similar problems, have similar interests, and aspire to similar objectives in the international system. These shared economic concerns and imperatives, it is argued, lend a measure of economic coherence to the Indian Ocean area. Another indication of that emerging economic coherence is to be found in the attempts at economic cooperation within various cores of the region.

This has been most evident in Southeast Asia where the states of ASEAN have achieved an impressive degree of economic cooperation since the formation of their organization in 1967. Despite numerous problems, the association has made a significant contribution to economic development in the area.

At the same time, the perceived advantages of association have given impetus to the settlement of some regional political issues such as the long-standing territorial dispute between Malaysia and Indonesia over the Malacca Strait. Another systematic attempt to form an integrated economic association, the East African Community, established by Kenya, Tanzania, and Uganda in 1965, has failed to achieve the expectations of its founders. Between 1971 and 1979, while Idi Amin was in power in Uganda, it virtually ceased to function.

While there are now signs that the founding members of the community are reviving their interest in it, differing economic ideologies, particularly evident between socialist-oriented Tanzania and free-market-inclined Kenya, could well prove to be an insurmountable impediment to the integration process.

A more recent experiment in economic cooperation is the Gulf Cooperation Council formed in the spring of 1981. Consisting of Bahrain, Kuwait, Oman, Qatar, Saudi Arabia, and the United Arab Emirates, the GCC has established administrative headquarters in Riyadh and is touted as a first step toward the economic, social, political, and military integration of member states.

The GCC is being built on a pattern of cooperation among Persian Gulf states which emerged during the 1970s, largely as a result of British withdrawal. Such cooperation has taken the form of jointly owned and operated airline and shipping companies, multilateral financial aid institutions, trade liberalization agreements, and joint industrial and service ventures. When its formation was announced, heavy emphasis was placed on economic objectives as the raison d'être of the GCC's existence.

These economic aims were rapidly eclipsed, however, by the urgent security concerns created by the Iranian revolution and the Iraq-Iran war. When the security environment becomes less threatening, the GCC may be expected to emphasize its economic agenda once again. What role, if any, Iraq and Iran might ultimately play in the organization remains to be seen. A venture in expanded economic cooperation also has been launched in southern Africa.

In April 1980 the black states of southern Africa signed the Lusaka declaration to establish the Southern African Development Coordination Conference (SADCC). Among the stated aims of the conference were the reduction of economic dependence generally (and not merely with regard to South Africa) and the forging of links to create genuine, equitable, regional integration.

Neither of these objectives will be easy to achieve in the circumstances prevailing in southern Africa. The economies of the Frontline States (Angola, Botswana, Mozambique, Tanzania, Zambia, and Zimbabwe) rely heavily on trade with and access routes through South Africa. This dependent relation ship is one that the South African government will not wish to see changed and Pretoria can be expected to continue to pursue policies that will make the black states' tasks of reducing dependence and increasing economic cooperation among themselves extremely difficult.

In the longer term, however, a South Africa with a majority government would be a natural partner in any subregional economic organization. In the South Asia core, where instability and conflicts have hitherto foreclosed all avenues to closer relations, there is both potential for and an indication of growing economic cooperation.

The decline in tensions between India and Pakistan, partly as a result of the Soviet invasion of Afghanistan in December 1979, has given impetus to the expansion of the trade relations that had begun to develop between the two countries after the rapprochement of 1973.

Similarly, there has been an increase in trade relations between India and Bangladesh. India's economy will likely dominate any regional economic association in South Asia, a factor that clearly poses a major obstacle to cooperation. And Indo-Pakistani relations will undoubtedly continue to have their ups and downs. But the diversity of the Indian economy with its expanding industrial and manufacturing sectors complements others in the area and provides a basis for cooperation.

The sector of the Indian Ocean littoral where economic cooperation is largely nonexistent is on the Horn of Africa. Somalia and Ethiopia, the area's two principal states, are among its poorest. In recent years their wars against each other, together with a sustained period of drought and famine, have caused major economic disruption.

There appears to be little prospect that either of these blights will disappear from the horn in the near future. The various efforts at economic integration in the Indian Ocean have obviously had mixed success.

Yet it seems clear that the area's LDCS are aware of the need to accelerate their development and that they see economic cooperation as a useful vehicle for abetting the process. Although the more distant consequences are uncertain, it is likely that the evolution of regional economic organizations would improve the LDCS' trading performance, both within and outside the region. Such organizations might also prove to be useful instruments for the management of common resources such as fish and common problems such as pollution which overlap the jurisdictions of the extant subregional organizations.

Whether the scope of these organizations will broaden to match the scope of regionwide problems remains to be seen. The third dimension of economic activity bearing on the issue of cohesion in the Indian Ocean is the state of intraregional trade. Generally, developing countries provide poor markets for each other's exports. To ensure economic survival, they must rely on export markets in developed countries outside the region.

In turn, those countries have provided the industrial equipment and expertise necessary for development and the manufactured goods to help meet rising consumer demand. The predominant place of developed countries in the trade activity of LDCS in the Indian Ocean region is reflected in their profiles, with LDCS and CSOES still looking to developed countries outside the region as their principal trading partners.

Yet, despite the continuing importance of extra-regional trade, for most of these countries the value of intraregional trade flows has increased significantly over the past decade. Substantial increases have taken place in exports from Australia to Southeast Asia (especially Indonesia), South Asia (India and Pakistan) to the Persian Gulf (Saudi Arabia, Iran, and the United Arab Emirates), the Persian Gulf (Saudi Arabia) to Southeast Asia (Singapore and Indonesia), and Southeast Asia (Singapore) to South Asia (Pakistan). Other less significant gains have been recorded between the Persian Gulf (Bahrain) and East Africa (Kenya) and between South Asia (India) and Southeast Asia (Indonesia).

These increases in part reflect developments that have had a worldwide impact on trade, namely, several years of inflation and a substantial increase in the price of oil as a result of the activities of the Organization of Petroleum Exporting Countries (OPEC). Beyond these factors, however, the increases reflect changes taking place within the region: an acceleration in the pace of development in some LDCS; a concerted effort on the part of one industrialized country of the region, Australia, to expand its markets in the area; and a diversification in the economies of several of the larger states (such as India) which has improved levels of complementarity. However, the increase in intraregional trade is not uniform.

For example, countries of Southeast Asia and Australia (the Australasia core) still have only small export markets in Africa. Similarly, only a few African states have found extensive markets in the Persian Gulf core or in Australasia. While an array of factors specific to each case has contributed to this situation, the lack of complementarity among the economies of many of the countries has been an important underlying factor.

Given the continuing efforts toward subregional economic integration and the likelihood that intraregional trade will continue to expand as area economies diversify, it is likely that the trend toward greater economic interaction in the Indian Ocean will persist. At this stage, however, regionwide economic integration along the lines of that in Western Europe, or even the more modest core model of ASEAN, will be beyond the capability of most governments in the area.

The lack of complementarity in LDC economies, incompatible economic systems, and the persistence of widespread political instability all act as restraints on economic interaction. The economic life of the Indian Ocean will be characterized by diversity and a relatively low level of cohesion for the foreseeable future but there are unmistakable signs that the long-term trend is toward greater economic interaction. The concept of an Indian Ocean region is not without its critics. Barry Buzan is one of the more articulate: "As things stand, the attempt to conjure up an Indian Ocean region tends to detract more from understanding than it adds. The problems of omission and superficiality which arise from the scale and diversity of the area, are not offset by the weak and tentative linkage which the Indian Ocean framework provides."

This chapter has been an attempt to examine both theoretical and empirical bases for viewing the Indian Ocean as a regional subsystem of the global political system. The systems-analytical perspective has been suggested as a tool for coping with "the scale and diversity of the area," as a tool for bringing some structure to thinking, some order to description, and some inspiration to research.

Furthermore, it has been suggested that the conceptual abstraction of an Indian Ocean region has an empirical basis in the "real world" of indigenous and external policy makers, reflected both in their rhetoric and in their actions. Indeed, it has been argued that the empirical case for the concept of an Indian Ocean regional subsystem is becoming stronger over time. Buzan is right to inveigh against potential sins of omission and superficiality.

The Indian Ocean region is undeniably large and diverse but so too are Latin America, the Middle East, the North Atlantic community, the Commonwealth, the Far East, the Mediterranean, and other geographic and conceptual generalizations commonly used in international political analysis.

The key question here is whether the concept of an Indian Ocean region is a useful one, whether on balance it adds to or detracts from our understanding of political and strategic reality. The consensus among contributors to this volume clearly appears to be that the concept adds to understanding. Analysis from the perspective of an Indian Ocean region is not intended to replace the labors of country specialists or of traditional area specialists.

It is also not being touted as the best approach toward enlightenment on all the questions worth asking about the international politics of the Indian Ocean area. Rather, the regional perspective is meant to complement, supplement, and incorporate insights from more traditional analysis. The objective is the advancement of overall knowledge of an increasingly strategic area of the world.

2

Economic Developments In Indian Ocean

The Indian Ocean Rim Association for Regional Co-operation (IOR-ARC) was one of the latest initiatives in the establishment of regional arrangements for economic co-operation. This initiative is based on the observation that many new economic groupings are being conceived what is now called the "domino theory" of regionalism - and motivated by the conviction that the process of co-operation on a regional basis would be beneficial. IOR-ARC was formed as some Indian Ocean region countries wanted to assert their space as a response to other mega trading blocs, in particular, the North American Free Trade Agreement (NAFTA) and Asia-Pacific Economic Co-operation Forum (APEC); and the need of Australia, India, and South Africa to play a leadership role in the global market.

However, unlike most regional cooperation arrangements, IOR-ARC was formed on the basis of the APEC model where the concept of "open regionalism" is emphasized. Open regionalism involves regional economic integration without discrimination against economies outside the region.1 In contrast, discriminatory regionalism involves preferential trading arrangements, free trade areas, and customs unions, where official trade barriers for members are lower than for non-members. The concept of open regionalism emerged from, and has helped, the practice of economic co-operation in the Asia-Pacific region.

Open regionalism was articulated by the first Pacific Economic Cooperation Council (PECC) - the precursor to APEC - in 1980 as an ideal for the future development of economic relations in the Asia-Pacific region. APEC embraced the concept in 1989 at its first inaugural meeting in 1989. The second regional bloc to embrace this concept was the IOR-ARC. Unlike APEC, which took nearly three decades to materialize after the initial conception of Pacific economic co-operation and nearly a decade after the formation of PECC, the IOR-ARC came into operation without any serious attempt to engage in economic co-operation in the past.

An Indian Ocean Commission was formed by Mauritius and its neighbouring islands in 1984 for economic co-operation but it hardly managed to take off and is currently very close to the stage of natural disintegration. The IOR-ARC was officially formed in March 1997 with 14 member countries, namely, Australia, India, Indonesia, Kenya, Madagascar, Malaysia, Mauritius, Mozambique, Oman, Singapore, South Africa, Sri Lanka, Tanzania, and Yemen. Since this regional grouping covers a vast geographic area, different cultures, large distances, and involves countries at very different stages of trade liberalization, it has been questioned whether economic co-operation in the IOR-ARC under the open regionalism framework would be meaningful in a practical sense.

This chapter attempts to shed some light on this question. The idea of forming an Indian Ocean Rim (IOR) trading bloc was first mooted by the then South African Foreign Minister, Pik Botha in November 1993 during a visit to India. This idea gathered momentum after the subsequent visit of the South African President to India in January 1995. Consequently an Indian Ocean Rim Initiative (IORI) was formed by South Africa and India, with the assistance of Australia and Mauritius. It started with the concept of economic cooperation, and with the following broad objectives:

(1) to stimulate intra-regional trade and investment;

(2) to synergize competitive advantages in commodities, manufactures and services;

(3) to collect, classify and distribute data and information;

(4) to help establish a network among IOR countries; and

(5) to promote standardization and harmonization of data, statistics, procedures, etc. (IOC various issues).

The first inter-governmental meeting resulting from the IORI to discuss the formation of IORARC took place in end-March 1995 in Mauritius with the participation of the four earlier mentioned countries together with Kenya, Singapore and Oman.

The discussions in this Mauritius-initiated meeting were based on the APEC model, where business and academic groups participated with government officials. In other words, the discussions were tripartite in nature and focused solely on economic issues. Political and strategic issues were deliberately excluded, at least at the inter-governmental level.

A Working Group comprising the seven states were formed to formulate a Charter and a future Work Programme. Subsequently, the Indian Ocean Rim Academic Group (IORAG) and Indian Ocean Rim Business Forum (IORBF) were formed. IORAG and IORBF were supposed to work together to identify special areas of economic co-operation and submit the proposals to the government officials for approval. Most projects under these two bodies were supposed to be funded by the respective governments.

Meanwhile, Australia promoted an International Forum for Indian Ocean Region (IFIOR) in mid1995 where

23 states from the IOR participated. This was a non-governmental meeting where government officials participated in an unofficial capacity together with academics and business groups. This initiative was called the "second track" of the IOR and here too, an academic group called the Indian Ocean Research Network (IORN) and a business group called the Indian Ocean Rim Consultative Business Network (IORCBN) was formed. The groups were supposed to find funds from independent sources and the private sector that had interests in Indian Ocean affairs for the approved projects.

The Working Group of the Mauritius initiative, with IORAG and IORBF, met on three occasions after the initial meeting in March 1995 to prepare the Charter and the Work Programme on ten specific projects. With this background work, in March 1997, IOR-ARC was formally launched with seven additional countries as members. They were: Indonesia, Sri Lanka, Malaysia, Yemen, Tanzania, Madagascar, and Mozambique. The IORAG and IORBN followed up the ten projects and took a decision to submit the final reports to the inter-governmental meeting held in Mozambique in 1999.

The Mauritius initiative, which is now known as the "first track", kept a distance from the Australian initiative of the "second track" and by mid-1997 it was clear that the "second track" had almost reached a point of disintegration. At present, when one refers to Indian Ocean regional co-operation, the reference is to the "first track" or IOR-ARC - the official inter-state arrangement of regional co-operation. The focus in this chapter will be on the 14 member countries of IOR-ARC.

In contrast to APEC, where economic cooperation is market driven, the IOR-ARC economic co-operation had been driven by the initiative of respective governments from the start. As discussed earlier, the IOR-ARC was formed as a response to growing regional trading arrangements by some of the IOR countries in order to assert their regional space. It was the government officials who decided to include the business and academic segments in IOR-ARC affairs.

In contrast, it was the business and academic segments in the PECC that put pressure on Asia-Pacific governments to form APEC. The operational mechanism of the IOR-ARC would be as follows: A committee of senior officials of the FOR-ARC (composed of government officials of member states) will review implementation of decisions taken by the Council of Ministers on the basis of suggestions made by IORAG and IORBF.

In consultation with IORAG and IORBF, the Senior Officials will establish priorities for economic co-operation, mobilize financial resources for the Work Programmes, monitor and co-ordinate them. Intra-regional trade is a good indicator of the existing level of economic co-operation. In 1995, IOR-ARC intra-regional trade as a percentage of world trade amounted to nearly 19 per cent.

This level of intra-regional trade is almost close to the level of intra-regional trade in ASEAN in 1991. Panchamukhi estimated IOR-ARC intra-regional trade at 14.9 per cent in 1993. Clearly, over a period of two years intraregional trade has increased remarkably, that is by 4.1 percentage points. A disaggregated analysis shows that IOR-ARC intra-regional exports amounted to 20.3 per cent of IOR-ARC's world exports, and intra-regional imports amounted to nearly 17 per cent of IOR-ARC's world imports in 1995.

Intra-regional trade varies from a high value of 44 per cent for Mozambique, to an average figure of 21 per cent for Singapore to a low value of 6.4 per cent for South Africa. Given the fact that intra-regional trade among ASEAN countries, at the start, around 1970, was about 6 per cent, these figures for IOR-ARC appear to provide a good starting point for economic co-operation. In regard to investment promotion, all IORARC countries welcome foreign direct investment (FDI) from within and outside the region.

The amount of intra-IOR-ARC FDI flowing from one country to another, especially from Malaysia, Singapore, Australia, India, and Indonesia, has been increasing. For countries like Australia and Singapore where there is a need to offset their growing comparative disadvantages due to higher wages and operating costs, FDI has become an important instrument for conducting business in certain areas.

The large existing and potential size of the home market in some IOR-ARC countries provides the setting for attracting FDI from more industrialized IOR-ARC countries as well from outside. From both the trade and investment areas, prima facie, it appears that IOR-ARC has a good potential for economic co-operation. But is this really the case? To answer this it is essential to comprehend the fundamentals of open regionalism.

ANALYTIC BASIS OF OPEN REGIONALISM

In the literature, there is considerable and lively debate on the concept of open regionalism. The critiques have found the term inherently contradictory, arguing that arrangements that are open cannot be regionally confined, and those that are regionally confined cannot be open, and Prof. T.N. Sirinivasan, the most vocal critic of the idea, has gone to the extent of calling open regionalism an oxymoron. It is not the intention here to examine the pros and cons of open regionalism but to discuss the concept as put forward by its proponents such as Garnaut, Bergsten, and others, and thereafter discuss its relevance to IOR-ARC.

According to the proponents of open regionalism, to comprehend how open regionalism stimulates economic integration one has to first address the nature of resistance to trade, and methods for overcoming it. Then it is essential to understand the concept of market integration and the factors that facilitate market integration. The normal state of world markets is one of considerable lack of integration: wide price disparities between regions, between countries, and within countries. Lack of integration persists because of barriers, or resistance to trade. Resistance to trade is defined as a phenomenon which prevents or retards the movement of commodities in response to price differentials.

One line of literature that has developed out of the analysis of Asia-Pacific experience distinguishes two basic types of resistance: objective and subjective. Objective resistance can be overcome by firms only at some objectively determined minimum cost. They comprise principally costs of transport and communication, costs of overcoming distance, and official barriers to trade (principally protection).

Subjective resistance comprise a range of social, physiological and institutional factors which cause prices to vary across geographic space by larger margins than can be explained by the necessary cost of overcoming objective resistance. Subjective resistance derives from perceptions of risks and uncertainty about property rights and valuations at various stages of trade transactions, from imperfection of information available to firms, and from the process through which firms make decisions that affect the volume, geographic direction or commodity composition of trade.

Economies of scale affect the cost of overcoming resistance such as objective transport and communication costs and subjective resistance of all kind, but not official resistance. The former resistance include:

    (1) externalities stemming from one firm's investment to reduce the cost of overcoming resistance to trade;

    (2) investment in the organization of a new pattern of transport or communication that could reduce cost for other firms in the trading relationship; and

    (3) investment in information to support a new pattern of trade which provides information to others including through observation of the resulting trade expansion itself.

The presence of externalities in investment, to reduce subjective resistance to trade, and to bring transport and communication costs closer to their objective minima, requires the possibility of economically efficient roles of government, quite separate from those associated with the imposition of official barriers to trade. These are the roles of the governments in providing public goods that are relevant to efficient operation of the international market:

    (1) improving transport and communication infrastructures;

    (2) improving regulatory regimes;

    (3) reducing perception of risks in international contracts; and

    (4) dissemination of information of profitable trade opportunities.

So where private institutions are inadequate in internalizing externalities, there can be gains from provision by governments of "public goods" required to maintain a high level of trade. The reduction of resistance takes investment and time, and is affected by a whole range of cultural, linguistic, legal and other factors that affect the costs of trade transactions. Much of the dynamism in Asia-Pacific trade expansion derives from the progressive reduction of subjective and non-official objective resistance to trade.

The process has been driven by independent enterprises' search for more profitable patterns of trade, sometimes assisted by governments' provision of "public goods" that affect the operation of private markets. In other words, the process is market driven and leads to "market integration" and works without preferential policies.6 Integration can occur as a result of governments getting out of the way of profit maximizing patterns of trade; or through dynamics of private discoveries of profit maximizing investment, without any change in the policy stance of governments. Market integration is likely to intensify intra-industry trade.

How can market integration be facilitated? Achievement of the full potential of market integration requires low official barriers, and some inter-governmental support for exchange of goods and services.

Market integration involves reduction in subjective resistance to trade, but builds upon low objective resistance. Thus geographical proximity is important for two reasons. First, it introduces the potential for low transport and communication costs. Second, it facilitates reduction of subjective resistance due to easy human contact.

Cultural and language similarity also facilitates the reduction of trade resistance through the market process. What is open regionalism? "Open regionalism" includes market integration and also market integration that is facilitated by government policy to the extent that it does not involve discrimination against outsiders.

Thus, unlike in the conventional text book type of regional economic integration — where the grouping goes through the stages of a Preferential Trade Arrangement, to a Free Trade Area, to a Customs Union, to a Common Market, and finally to an Economic Union - regional groupings under "open regionalism" go through the market process and non-discriminatory market facilitation process by the state.

There are three ways in which open regionalism could stimulate economic integration. First, through non-discriminatory reduction of protection. Second, the expanded provision by the governments of public goods relevant to the efficient operation of regional or international markets for goods and services, but without any element of discrimination in official barriers.

Third, through the process of market integration which is the major source of dynamism in open regionalism? Market integration is the purest form of open regionalism since by its very nature it contains no element of de jure or de facto official discrimination.

IOR-ARC AND OPEN REGIONALISM

Using the above conceptual framework, we could start by first asking the question: what is the resistance to trade in the IOR-ARC? It is very clear at the onset that non-official objective resistance is high in the IOR-ARC because of high transport and communication costs resulting from the vast area covered by the Indian Ocean region and the lack of commonality between some members; for example, there are no direct flights between many of them and air travel involves transits and delays.

Subjective resistance is also at a high level because of the following factors: the heterogeneity of members; the fact that they have not co-operated at an unofficial level for nearly a decade as in the PECC; perceptions of risks; uncertainty about property rights; imperfect information; the fact that the private sector in IOR-ARC nations is not developed enough (other than in Australia and Singapore, and to some extent in Malaysia, India, and South Africa) to reduce subjective resistance in the short run. So irrespective of trade protectionist barriers, both objective and subjective resistance to trade are high. Thus the member governments have to play a role in providing these "public goods".

The provision of public goods: The roles of the member governments are less dynamic than in APEC countries in the provision of the "public goods" to overcome subjective resistance. This is clear from the funding and the progress of the projects that were agreed on in the IOR-ARC meetings. Most member countries have not provided adequate funds for these projects and there is little evidence of any noticeable progress of these projects. Garnaut argues that "open regionalism involves commitment to high level of provision of institutions to support intraregional market exchange [exchange of goods and services]".

This seems to be lacking in the IOR-ARC. An Australian suggestion during IFIOR to form an Indian Ocean Economic Cooperation Council (IOECC) along the lines of PECC (that was instrumental in forming the APEC forum) was not viewed favourably by some members of the IOR. These members took the view that new institutional arrangements in the region, where subregional organizations already took up a lot of time, should not be a priority. Establishing a Secretariat in Mauritius was not regarded favourably at the first Ministerial Meeting of the IOR-ARC in March 1997.

Some members were of the view that "there should not be too much bureaucracy in the set up". In short, all attempts to have institutional mechanisms in the IOR from the very start have faced problems. Clearly, the member states still do not seem to be at ease with one another. Market integration involves developing low objective resistance, and reducing subjective resistance. Clearly, since objective resistance is high, and progress in reducing subjective resistance is slow, there appears to be less scope for market integration among IOR-ARC countries at present. Moreover, market integration within IOR cannot draw on geographical proximity and cultural and language similarity.

Some studies have attempted to argue that market-driven co-operation already exists among IOR countries, in particular, India, Australia, and South Africa, which could spread to other countries. This argument is based on increasing intra-industry trade over time, measured using the Grubel-Lloyd index, for three-digit SITC-level industry. But even if intra-industry trade is increasing, it is still at insignificant levels as a proportion of overall trade.

In 1994/95, Australia's share of South Africa's exports was only 0.75 per cent, while Australia's share of South Africa's imports was only 0.65 per cent. In 1992/93, India accounted for 1.5 per cent of Australia's total exports while Australia received 0.6 per cent of India's exports. These figures have not significantly changed by 1995. Moreover, India's bilateral trade with South Africa is hardly impressive. Thus market-driven trade co-operation appears to be insignificant at least among the three major poles of IOR-ARC. Is there any evidence of significant intra-IORARC investment? Data is not readily available to make a proper analysis.

But from the available scattered data for India at least there does not appear to be significant intra-IOR-ARC investment taking place. For example, IOR-ARC investment in India in 1994 was U.S.$284.3 million, which was only 17 per cent of total FDI in India in the year 1994. On the other hand, Indian investment in IOR-ARC countries was a meagre U.S.$90 million in 1995 compared to the total FDI in IOR-ARC countries of US$181,433 million in 1995 - 0.5 per cent of total FDI in IOR-ARC.

After the East Asian currency crisis there were signs of Australian investment flowing to India, but it is unlikely after India's nuclear explosion in mid-May 1998 given Australia's strident support for U.S. sanctions on India. The East Asian crisis will also slow down the sustained growth of IORARC-member East Asian countries spilling over to other IOR-ARC countries in the form of an increase in trade and investment? Thus the open regionalism that is associated with IOR-ARC is not well supported by market integration or government policy. Is it supported by non-discriminatory liberalization?

The nondiscriminatory reduction of protection is taking place at a rapid pace in some IOR-ARC countries that are currently members of APEC according to the APEC time-table. In other IOR-ARC countries, reduction of protection is taking place via the ongoing liberalization policies, but at a much slower place than in APEC. This is because, first, these IOR-ARC countries are at very different stages of liberalization compared to APEC countries. For example, Madagascar and Mozambique can hardly be considered open economies, and may take more time than other members to fully open up their economies. Second, as in APEC's Bogor Declaration, there is no formal agenda to reduce tariffs to zero by the year 2010 by developed countries, and 2020 by developing countries.

In the context of APEC, Yamazawa states:

... the real test of the success or failure of the Osaka APEC meeting is not the existence or absence of a formal agreement, but whether business people are convinced that member governments in the Asia-Pacific region will steadily improve the business environment.

In the IOR-ARC, the business community's perception could be gauged by IOR second track activities just as the business communities' enthusiasm was gauged in the PECC before the formation of APEC. Both the IORBCN and IORN met in Delhi in December 1995 and in Durban in March 1997, after the first meeting in Perth. In both meetings there was no strong indication of the selected projects being funded by independent sources, such as the private sector.

With the exception perhaps of the Chambers of Commerce and academic institutions in India and Australia, no other member-country institutions committed significant funds for IOR second track projects. Even if the IORN failed to mobilize independent support for their projects, it is difficult to comprehend why the business grouping, IORCBN, failed to mobilize funds for their projects, if a driving force existed for more market-driven integration. Clearly, the private sector in member countries still seems to be ambivalent about the future prospects of the IOR-ARC.

Once open regionalism starts operating it creates its own problems. APEC's experience sheds light on some of these issues, such as:

> (1) problems of co-ordination and operational efficiency due to its relatively unstructured nature, given the low emphasis on institutions;
>
> (2) reconciliation of "open regionalism" in APEC with "closed regionalism" of NAFTA and ASEAN;
>
> (3) the free rider problem (APEC goal of full liberalization by the target date will open up their economies to non-member countries who have not fully liberalized their economies);

(4) vagueness of the individual action plans (IAP) for liberalization. Though the IOR-ARC has not experienced any of these yet, whether IOR-ARC can handle such problems in the future in the way APEC does at present remains questionable, given the diversity and different levels of development among the member countries.

IMPEDIMENTS TO ECONOMIC CO-OPERATION IN THE IOR-ARC

Five major impediments for a strong IOR-ARC can be identified. First, integration choices are not merely a matter of economics, but are very much related to domestic and regional politics. The EEC came into being, inter alia, because there was a perception that Europe was lagging behind the United States and Japan in the 1960s, and that one way to face this challenge was through economic integration in Europe. ASEAN came into being, inter alia, to face the security threat of growing communism in Southeast Asian countries such as Cambodia, Laos, and Vietnam. United States isolation from existing trading blocs, the growing power of the EC after the collapse of communism in Eastern Europe, and the increasing economic power of Japan in the 1980s provided the basis for the formation of NAFTA.

In the case of IOR-ARC there does not seem to be such a driving force behind its formation, and it seems to be more a reaction to the flurry of regional arrangements that came into force in the 1990s, particularly APEC and NAFTA. South Africa had just emerged as a free nation and was attempting to assert its existence in a wider regional space, India was not very satisfied with SAARC (South Asia Association of Regional Cooperation) and was seeking wider economic opportunities to support its liberalization policies initiated in 1991, and Australia was trying to cover up its isolation in regional blocs by new regional arrangements concordant with its "Look West" policy.

The response to these individual factors was the formation of IOR-ARC with 11 more countries which had little similarities. Merely wanting to assert their regional space in the absence of a security threat, or in the absence of the need to compete with another regional bloc simply does not provide an adequate driving force for co-operation, even if the co-operation is exclusively for economic matters. Second, the common denominator of sharing the Indian Ocean waters by itself does not provide an adequate economic rationale for regional cooperation. There is no clear link between the degree of heterogeneity of the countries included in a regional co-operation grouping and the intensity of co-operation among them.

The general experience is that the most successful regional groupings have been those where the region as a whole tends to grow together - for example, EC and ASEAN because countries are more or less at a similar level of development and subject to similar business cycles. This is not the case among IOR-ARC countries. One can argue that the countries in APEC are as heterogeneous as in the IOR-ARC. However, all APEC member countries have achieved a certain level of industrialization; and notably, the inter-industry trade was already at a relatively high level in 1990.

In addition, most member countries are also members of one or other of a dynamic regional bloc, such as ASEAN and NAFTA. Thus, the institutional links are already in place for these countries to come together successfully under a large umbrella such as APEC.

As in APEC, some IOR-ARC countries are already members of other regional groupings such as ASEAN, APEC, SAARC, IOC (Indian Ocean Commission), GCC (Gulf Co-operation Council), SADC (South Africa Development Community), EAC (East African Co-operation), COMESA (Common Market for Eastern and Southern Africa), and CBI (Cross Border Initiative). But unlike APEC, all members of ASEAN, SADC or SAARC are not members of IOR-ARC. Thus it is more difficult to tread a complementary path without coming up against a conflict of interest between subregional groups."

The overlapping membership of regional groupings, and the conflicts mentioned above, dilutes the effectiveness of, and commitment to a larger and less cohesive group such as IOR-ARC. It is unlikely that IORARC members would get together in international fora to strengthen their bargaining position, because of their weak commitment to this particular group and conflicting interests. Third, some IOR-ARC countries such as India and Sri Lanka which do not have any Pacific boundary have already applied for membership in APEC, and India applied for membership in NAFTA in 1995. Thus it appears that membership of IOR-ARC for at least some countries are sort of a "second best" option in the absence of membership in a key trading bloc.

This is not to say that the present commitment of these countries towards IOR-ARC is diluted, but that they are looking beyond IORARC, and in the event these countries succeed in their attempts, the commitments may be diluted. It is argued by some commentators on IORARC that countries such as Thailand, Iran, Egypt, Bangladesh, Pakistan and Seychelles have applied for membership of IOR-ARC because of its future potential. In this context, it is important to note that no country likes to be isolated from a regional bloc oriented towards economic co-operation, specially when neighbouring countries are members of that regional bloc.

There is always a tendency to avoid isolation and secure benefits from perceived economic potential. Thus the fact that some countries are on the waiting list to obtain membership does not by any means indicate that there is a great potential for economic co-operation in the IOR-ARC. Fourth, in ASEAN, besides factors such as cultural and linguistic similarity and geographical proximity, direct foreign investment from Japan, and later from Taiwan and Korea, especially from the late 1980s, facilitated trade expansion, and the beginning of intra-industry specialization in manufactured goods production.

In the IOR-ARC, in the first place, there is no major non-member investor in the region; second, only two member countries have the potential to make substantial investments in other countries, namely Singapore and Australia (and perhaps Malaysia and India). Singapore is committed to APEC and most of its investment is likely to be channelled to the APEC region. Thus the only country with investment potential seems to be Australia. But can Australia play the same role in IORARC as Japan played in ASEAN? This is very unlikely.

Fifth, how viable are sectoral co-operation projects in IOR-ARC? Panchamuki argues strongly for sectoral co-operation, and has identified several sectors for possible co-operation in the IOR-ARC. It is argued that IOR will make sectoral co-operation more viable because IOR-ARC will provide a formal forum for implementation, and set formalized institutionalized linkages among member countries. Such sectoral co-operation can lead to a more complex regional trade arrangement if significant mutual benefits can be gained, and if the member countries are dedicated to the principle.

This may perhaps shape the future economic co-operation structure in the IOR-ARC. However, the sectoral approach to economic integration of the IOR will be a time consuming process, since it involves consensus among all member countries, and consensus among the academic, business, and officials of all IOR-ARC member states to identify key areas for co-operation that would be advantageous to the whole region. An attempt at developing sectoral co-operation in the Indian Ocean Commission was not successful.

McDougall observes: "... disparity [among countries] means that the interest of the members is often different. Consequently the range of regional co-operation projects which can be attempted is limited." Given the diversity among countries, this appears to be the case for IOR-ARC too. Evans argues that in ancient times seaborne commerce in the Indian Ocean had generated a thriving network of trade and peopleto-people links. But with the arrival of the Europeans in the fifteenth century, the Indian Ocean economies, which were earlier tightly interdependent, were restructured to meet extraregional imperatives.

And this erosion of regional cohesion continued even after the European presence faded away from IOR countries in the post-Second World War years. But now the environment is conducive to resurrect the past glory of the Indian Ocean. Singh has pointed out that IOR is endowed with a large reservoir of energy resources, valuable minerals, and so on, that form the backbone of the modern development process.

IOR ... accounts for about 40 per cent of world's gold, 90 per cent of diamond and 60 per cent of uranium. At the moment, most of this wealth remains either unexploited or traded as raw material and does not fully contribute to the development of the region. Based on such sentiments and aggregate statistics, these commentators have argued that there exists a significant potential to reap benefits from IOR economic co-operation.

In cases such as IOR-ARC it is best to look at the issue of economic integration from the perspective of the type of regional framework governing it - open regionalism, and not from the current levels of intra-regional trade, or for that matter, from the number of applicants in the waiting list seeking membership. It is essential to analyse whether IOR-ARC satisfies some of the conditions that make an open regionalism arrangement work. This chapter showed that, at least at present, IOR-ARC does not satisfy most of the modalities of the open regionalism framework where market integration is the driving force.

Furthermore, statistical evidence on current economic links must be combined with political economy factors governing the regional grouping to make any meaningful conclusion about the feasibility of economic co-operation. In the IORARC, the political factors favouring stronger economic integration are rather weak. The fact that political-economy factors are important in economic co-operation is demonstrated by the failure of previous attempts in the IOR. As is well known, co-operation among Indian Ocean region countries is not a new phenomenon, and there have been several such attempts in the past.

At the peak of the Cold War in 1971 Sri Lanka took the initiative to form the Indian Ocean Peace Zone (IOPZ) to create a climate of peace and security in the region, free of great power rivalry, but it failed to take off. As stated, the Indian Ocean Commission which was formed in 1984 has now run out of steam and is very close to a natural death. In 1987, the Indian Ocean Marine Affairs Cooperation (IOMAC) was formed for co-operation and management of the ocean and its resources, but the body is yet to be formalized as the progress of its activities has been abysmally slow.

The primary reason for the failure of most of these arrangements or groupings has been the lack of political will and commitment. Given the limited domestic support in most member countries for regional co-operation in the Indian Ocean, IOR-ARC related projects have been given low priority by the IOR-ARC policy makers during the last few years, To the extent that the domestic policy support remains limited, the decision makers' regional co-operation initiatives will continue to be ambiguous, sporadic, and fragmented, leading to a "stop and go" pattern of regional co-operation in the IOR.

In such a pattern of co-operation, IOR-ARC growth in terms of regional institutional developments and programme implementation will remain uncertain and the organization's life cycle will oscillate between short-lived euphoria and agonizingly protracted stalemates. A firm commitment is lacking at present in the IOR-ARC for meaningful progress towards economic co-operation.

Given this reality, IORARC at this stage should not be too ambitious and try to emulate the APEC model, because it does not possess the same foundation that APEC had at its inception. Thus it could ill afford, for example, to emulate APEC's time-table for trade liberalization. IOR-ARC has a long way to go to achieve some meaningful economic integration. Meanwhile, IOR-ARC should become a forum to build up the spirit of economic co-operation and induce a commitment by member nations to implement a dialogue to promote a regional environment conducive for growth in trade and investment. Such a move would lay the foundation for more ambitious economic activities in the IOR-ARC in the future.

3

Andaman & Nicobar Island, Lakshadweep: An Extension For Military Capability

The Andaman Islands are a group of islands in the Bay of Bengal, and are part of the Andaman and Nicobar Islands Union Territory of India. Port Blair is the chief community on the islands, and the administrative center of the Union Territory. The Andaman Islands form a single administrative district within the Union Territory, the Andaman district (the Nicobar district was separated and established as a new district in 1974). The population of the Andamans was 314,084 in 2001. Andaman Islands - Physical Geography.

The factors that could spark conflicts in the 21st century, the international and regional strategic environment, and the dimensions of possible conflicts, are aspects that need the focus of the Indian government and the strategic community in India, in order to assess their impact on the nation's security, economy and the social structure. The connotations of demographic movements occasioned by the pressures of the population explosion in the neighbouring countries, is already an element that India has had to deal with in the last few decades, in context of the fact that it is host to a few million refugees.

However, considering that the population explosion within India itself in the next 50 years, is likely to make it the world's most populated country, it would be stating the obvious that any additions to these numbers by migration from adjoining countries, will cause serious social tensions, economic upheaval, and environmental disaster. The challenge will, therefore, be to put in place appropriate structures to ensure that any migratory movements that may take place are part of mechanisms that are designed to absorb any adverse impact.

The only framework that would appear to lend itself to successful management of this challenge, is the institution of a South Asian Union on the lines of the European Union, with open borders and free trade, economic cohesion, including a common currency, and a cooperative political arrangement that is also answerable to the people of the region as whole. The feasibility of such an option lies in the common strains of ethnicity, culture, tradition, and aspirations of the peoples of the region. Needless to say, for such an arrangement to come into being, a very high order of statesmanship, determination, sagacity, and compromise, are required.

India, with its size, geographic location, manpower and material resources, large industrial base, technical expertise, and well-established democratic traditions, will need to be the driving force. The developed world, particularly the U.S., Europe and possibly Japan, could act as a catalyst in this remarkable venture to make it happen. The assistance that would be required is not in the form of doles (with the inevitable strings attached), but an infusion of investment particularly in the infrastructure sector of the South Asian countries, ready access to advanced technology in industry and agriculture, and more particularly in the exploitation of the renewable energy resources like solar energy, bio-technology, and the ocean bed.

A tall order on all counts, but not impossible. Should such an arrangement come about, it would also provide a credible and effective apparatus for the security of the region from external conflict influences, with considerably reduced demands on the countries of the region for allocations for individual defence needs. India's dependence on oil imports to meet its energy needs, and the fact that two oil rich areas, namely West Asia and Central Asia are in its proximity, has serious strategic implications for the country. Any conflict situation in either or both these areas would almost inevitably have a fall-out that will affect India.

Firstly, the flow of oil could be stopped or curtailed, thus severely affecting every sphere of activity in the country. Secondly, the country could get drawn into the conflict, should it get enlarged. India has very close traditional and cultural links with many of the countries of both regions. The primary requirement in this context therefore, is for India to exploit its own oil resources to the extent feasible, but more importantly, to reduce its dependence on fossil fuels, and exploit other abundant renewable sources of energy.

In a country like India, the scope for extensive use of solar energy is limited only by the degree of determination to harness it. Similarly, the application of biotechnology to generate energy has equally extensive scope; the fact that large sections of India's rural population are still in the 'bullock cart' age, may not be a bad thing after all. Harnessing the waters of the many great rivers that traverse the subcontinent, for the generation of hydroelectric power, is another area that will need more attention.

India's capacity to generate nuclear power is established; it is only restricted by the technology control regimes imposed on it by the western world. Once the dependence on oil as the primary energy source is removed, the scope for the subcontinent to get drawn into any conflict scenario in West Asia or Central Asia is substantially reduced. Even so, it would be strategically prudent to institute arrangements diplomatically and commercially, with countries of West Asia, with whom India has always had excellent relations, to deal with crisis situations that may arise.

Similarly, traditional links with countries of Central Asia like Kazakhstan and Turkmenistan should be exploited. While striving to take its rightful place in the international community by giving the right thrust and direction to its internal and external economic and commercial policies, and concurrently ensuring defence preparedness meets her national security concerns, India needs to set her diplomatic sights on affiliations and alignments that will deal with the international strategic environment.

All the right reasons exist for durable and mutually satisfying alignments with the U.S. and Europe; provided there is an understanding that India is to be an equal partner, there is no reason why security arrangements cannot be entered into whereby regional or international threats to peace and security are met jointly. India's traditional and time-tested links with Russia must be nurtured, and in fact strengthened; both countries need to be equal partners (together with others, if necessary) in the diplomatic battles towards ensuring a polycentric world order.

In looking for other global and regional affiliations and alignments, it is essential that India shed her propensity for posturing. Japan, Vietnam, and countries like Myanmar, Thailand, Cambodia, and Malaysia, are natural allies, subject to our ability to garner their support on matters that concern international affairs. All of them, with others in East Asia and South East Asia, as also Australia, are concerned about the emergence of China as a potential super power that could flex its muscles to pursue policies towards total domination of the region.

Hence, while pursuing a policy of engagement with the Peoples Republic of China, and seeking a solution to the border dispute, alignments with other countries in the region must be strengthened. Similarly, close relations, including regional security arrangements, with the Central Asian Republics should be sought and, if brought to fruition, consolidated. Together with Iran, this could prove a decisive factor in stabilising the region. In West Asia, it would be to India's advantage to establish and nurture close links with Israel.

On the African continent, there are many countries with which India has strong links; the best arrangement would be to strike alignments with the regional organisations that have been set up. South Africa must be recognised as a significant partner in this venture. Needless to say, all these alignments and affiliations, or security links, would carry greater conviction and credibility if the South Asian region acted as one entity; that is the hope that will secure for the region the ability to concentrate on efforts directed towards the well being of its peoples.

MILITARY PREPAREDNESS

In context of the global and regional environment and the national security responses discussed in preceding paragraphs, appropriate strategic measures and military preparedness need to be put in place. A viable military strategy in the current context should necessarily be one of deterrence, both conventional as well as nuclear, based on a credible military capability. The aim being prevention of war or adventurism by an adversary.

Such a military capability must allow the Indian political and military establishment the option of waging a war that may be forced upon us by an adversary on one front, while ensuring a credible defensive capability on a second front. If a war is forced upon us, our armed forces should be able to prosecute it in such a manner as to achieve pre-determined objectives bearing in mind the need to keep the conflict below the nuclear threshold. Prosecution of a short and intense war, limited in scope and extent, would appear to be the option to be exercised.

In doing so, the requirement to contain the internal dimension of terrorist attacks against military lines of communication, logistics infrastructure and the civilian population in order to create an atmosphere of chaos and confusion, will need to be catered for. This can best be done by the deployment of Central Police Forces and State Armed Police. Needless to say, they must be equipped, trained and prepared for the purpose. In an extreme scenario it may well be necessary to also deploy elements of para-military forces like the Rashtriya Rifles, the Assam Rifles and the Border Security Force.

There is a general perception within some sections of the strategic community in India that it has for too long been defensive and reactive in its approach to national security issues, and that it needs to look for and exercise more proactive options. This is an aspect that may well be receiving greater attention. To this end, as also to project a credible military capability, in addition to the normal process of military modernisation, a number of measures have been instituted in the recent past or are being considered for implementation.

The first relates to the process of jointness within the armed forces by the creation of the Headquarters Integrated Defence Staff that has been effectively implemented at various levels. It is to be hoped that the Government will initiate early action for the appointment of a Chief of the Defence Staff. Secondly, there is a concerted effort to focus on the long neglected maritime dimension of our national security particularly in context of the security of maritime traffic in the Indian Ocean. The establishment of a tri-Service regional headquarters in the Andaman and Nicobar Islands is not only a step in the jointness process but a major step forward in the enhancement of India's maritime capability.

Thirdly, the establishment of the Strategic Command is not only another move towards greater jointness but also a recognition of the vital importance of the effective management and control of our nuclear assets. Fourthly, among other assets being acquired or improved upon, are airborne early warning and control systems that will enhance strategic surveillance and monitoring capability, unarmed aerial vehicles, tactical and strategic missile and anti-missile systems, electronic warfare systems and so on. Fifthly, the long overdue requirement of strategic airlift capability is apparently being addressed together with added acquisitions and improvements for a credible strategic projection of air power.

Lastly, there is greater focus on increasing and improving the capability of Special Forces not only to provide a force multiplier option in the conduct of conventional operations, but also to enable effective action against terrorists. Such forces would also provide the ability to deploy rapid action forces at short notice in regional conflict situations where our assistance is requested. All this capability would, besides enhancing the military capability of the Indian Armed Forces to deal with operational situations that may arise, also allow for more effective and meaningful deployment for international peace operations when called upon to do so.

And finally, one would hope that with the indigenous capability in weapons and equipment production being established, and joint ventures being entered into with other countries in this field, India will shed its earlier reluctance and enter the arena of arms supplies to friendly foreign countries. Besides being a source of revenue, such a step would also be useful in terms of building strategic security relationships. India is seeking to expand its area of influence by developing security relations in all directions, especially so in South-East and East Asia, with a view to becoming a major player in the emerging balance of power in Asia.

Although India had initiated its 'Look East' policy in early 1990s, but assertive diplomatic endeavours were made after overtly declaring itself a nuclear weapon state in May 1998. The main reason cited then by the Indian Government for carrying out nuclear explosions was a threat to its security from nuclear China. India has used the same China card for pacifying the negative reactions over its nuclear tests from the other Asian countries, particularly those that have abjured any forum of a nuclear defence policy. It is also making efforts to establish strategic relations with a few of them such as Japan, Vietnam and Singapore.

There are apprehensions among the majority of the Asian countries about China's emergence as a major economic and military power in Asia. The outstanding disputes over territory and exclusive economic zones (EEZ) between China and some of its neighboring countries, particularly over the Spratlys Islands, present a potential of an armed conflict, though China's state policy advocates non-violent ways of conflict resolution. A nuclear India with a substantial military power, that includes a blue water navy, is trying to present itself as a counter weight to China and a factor of stability in the region.

The strategic understanding that is in the making between India and the United States reflects the U.S. willingness to accord India a role for becoming a proactive player in the Asian balance of power for checkmating China. The U.S. strategic partners and allies such as Japan, South Korea, Singapore and in future perhaps Vietnam too, are evolving a special relationship with India in conformity with the overall U.S. strategic interests in Asia. The common denominator for this informal, multilateral strategic relationship is the alleged threat from China.

This dimension of U.S.-Indo strategic partnership is evident from the recent enormous acquisitions of arms and military equipment by India from Russia and Israel and the U.S. silence over this huge military build-up, which will disturb the existing strategic balance in the region and would force other countries to indulge in an arms race. But this development has certainly increased the weight of India in the Asian security calculus.

Though India is strengthening its security related relations with many countries in South-East and East Asia, this chapter intends to focus primarily on India's emerging security ties with Japan and Vietnam, with which India has historical ties, and which figures prominently in the U.S. policy of containing China. How far will these evolving strategic relations go? That would depend on a number of internal and external factors, as Asia is experiencing a transitional phase in its evolving security and economic structures. The emerging relationship between U.S. and China would be the single most important determinant of Asian security scenario in the coming years.

INDIA'S LOOK EAST POLICY

As the Cold War ended and India lost the mainstay of its military support, the Soviet Union, New Delhi quickly moved to revitalizing its relations with the remaining sole Super Power, the U.S., and the Great Powers namely, Western Europe, Japan, China and Russia. At that time, Prime Minister Narasimha Rao initiated India's 'Look East' policies and visited South Korea and some of the ASEAN countries. By then, the Vietnamese troops had been withdrawn from Cambodia, which had been a major stumbling block for improving India's relations with ASEAN countries, since India had supported the presence of Vietnamese troops in Cambodia.

The basic underlying objective of India's new eastwards policy was to cash in on the newly emergent opportunities for maximizing room for maneuver. In order to expand its engagement, India introduced economic reforms to achieve higher growth rates and create some degree of compatibility with the existent liberal economic practices.

It was generally believed that if India continues to remain trapped in a moderate growth syndrome, China would emerge with a double-digit economic growth rate, as Asia's undisputed leader. Thus, there was a need to step up economic growth through extensive diplomatic and political engagements in terms of trade and foreign investments, with the faster growing economies in the world, particularly in Asia. On the security side, the end of the Cold War created an opportunity in the form of a super power vacuum in the region.

As both the U.S. and the Russia closed down their military bases, in the Philippines and Vietnam, respectively, India became active and sought to expand its influence in the region through bilateral relations as well as through multilateral frameworks such as ASEAN. In 1992, India became a 'Sectoral Dialogue Partner' of ASEAN. During the years 1992–1995, Prime Minister Narasimha Rao visited Thailand, Indonesia, Singapore and Vietnam.

The high level exchanges helped India secure a 'Full Dialogue Partner' status in 1996, that also allowed India entry into the ASEAN Regional Forum (ARF), where security related matters of the member countries are discussed. In the first half of the 1990s, Malaysia, Indonesia and Singapore took initiatives to establish security relationships with India on a reciprocal basis. Defence officials from these countries undertook visits to New Delhi for discussions on security matters.

The then Malaysian Defence Minister, Mr. Najib Abdul Razak, visited India and reached an agreement under which India was to assist Malaysia in strengthening its defence forces and in maintaining the aircrafts of the Royal Malaysian Air Force (RMAF), and the sale of fast patrol boats for the Royal Malaysian Navy (RMN). It was further reported that Indian experts would train RMAF pilots on MiG-29 aircraft. Also, an understanding was reached on using Indian expertise in marine commando training, coastal surveillance, anti-piracy operations, weather forecasting, coastal search and rescue operations, defence of ports and harbours and shallow water mining capabilities etc.

Another high-level delegation led by Indonesia's Chief of Naval Staff, Mohammad Arifin visited New Delhi to explore more effective ways for strengthening maritime cooperation between the two countries. Over the last ten years, India has been able to cultivate security related relationships with all the major countries of the ASEAN region such as, Indonesia, Malaysia, Thailand, Philippines, Vietnam, Singapore and Myanmar.

These relations cover a wide range of activities, starting from military sales, training, and maintenance of military equipment to satellite launches and cooperation in nuclear fields. Though substantial progress was achieved in developing relations with the ASEAN and the East Asian countries in the first half of 1990s, the pace could not be maintained due to political changes within India and the 1997-98 financial crisis in South East Asia. The 1998 Indian nuclear explosions created a negative impact on these relations between India, Japan and ASEAN countries.

THE NEW ENGAGEMENT

India began a fresh diplomatic offensive to minimize the adverse effects created by its nuclear explosions of 1998. This damage control exercise proved to be more successful than any one could have expected. India's gate-crashing into the nuclear club, not only was it gradually accepted, but also the major powers and the sole super power sought to maintain strategic relations with India. Former U.S. President Bill Clinton, Japanese Prime Minister Mori and Russian President Validimir Putin, visited India in turns and expressed desire for forging strategic partnerships with it.

Such response from the big powers had a pacifying impact on the ASEAN members, which had initially criticized the Indian nuclear tests. India, with its strengthening economic environment, emerging Information Technology (IT), industry, and enhanced strategic military capability, engaged the ASEAN and the East Asian countries with a new confidence.

It is with this recently assumed confident stature that India seeks to expand its influence and play a larger role in the security matters of Asia, which it intends to extend to the Pacific region as well. For such a role, the consent of main players in Asian security like the, U.S., Japan, Russia has already been accorded to India.

Presently, the Indians claim that, 'our area of interest extends from the North of the Arabian Sea to the South China Sea.' India's aspirations for domination and expanding influence are also reflected in the Annual Report 1998-99 of the Indian Ministry of Defence titled 'National Security Environment'.

It depicts a number of national security interests some of which are as follows:

1. 'To be able to effectively contribute towards regional and international stability.
2. To enable our country to exercise a degree of influence over the nations in our immediate neighborhood, to promote harmonious relationship in tune with our national interests.
3. To possess an effective out-of-the-country contingency capability to prevent destabilization of the small nations in our immediate neighbourhood that could have adverse security implications for us.'

These are clearly interventionist national objectives and a matter of grave concern for the smaller nations in India's neighbourhood, which have suffered in the past from Indian hegemonic designs.

Now, India is hell bent on even redefining its neighbourhood through coining new terms such as 'Southern Asia' which, in turn, enlarges the physical and geographical security parameters of what has so far been understood as the seven member region of South Asia and security parameters of South Asia.

India will increasingly cooperate with the major powers in Asia, in such strategic alliances, which can facilitate Indian hegemonic aspirations. In turn, India will demonstrably present itself as a counter-weight to China, which is a common concern of many countries in Asia, especially Japan.

INDO-JAPANESE SECURITY RELATIONS

While Tokyo is constrained by its World War II legacy to acquire the full range of military capability befitting an independent economic power, the Japanese see China as rapidly acquiring a power projection capability and modernizing its military forces, including a missile build-up that puts all of Japan within its range. Tokyo has been troubled by signs of increasing nationalism and allegedly growing assertiveness in Chinese policies in the region. According to Japanese sources, China is not willing to do away with the historical memories of the Japanese occupation during 1937-1945.

There is a growing anti-Japan sentiment in China due to these historic factors, in view of Japan's strengthening defence relations with the U.S., which is considered to be a principal source of threat to the vital Chinese national interests. Though Japan has security guarantees from the U.S., at times policy posturing of the U.S. in its relations with China, such as former U.S. President Bill Clinton's 'China first' policy, disturbed and lowered Japanese confidence. The speculation that the U.S. may reduce its military presence in the Asia-Pacific region further heightens Japanese security concerns. Tokyo is also concerned over the growing impression in Washington that Japan is incapable of resolving its economic and strategic problems to provide leadership, and become a counter-weight to China.

A large segment of Japanese society feels that Japan should amend its constitution and develop militarily to become a normal power. This situation forces Japan to rethink its strategic alliances and explore multiple military options for playing a leadership role in Asia. According to Japan's Defence Agency, the number of incursions by the Chinese naval vessels into Japan's EEZ and territorial waters has increased manifold. China denies any such incursions and maintains that the two countries have yet to formally demarcate maritime lines. China's territorial claims to the

Senkaku Islands in the East China Sea conflict with Japan's claims over them.

Tokyo wants a maritime boundary halfway between the country's two coastlines, but Beijing argues for a demarcation in line with the natural extension of the continental shelf, thus claiming territorial waters that almost reach the Japanese island of Okinawa, which is a military base for the stationing of U.S. troops, and being extensively used by the U.S. to carry out surveillance operations against China. The Chinese proposals for such a demarcation are unacceptable to Japan. In 1997, the Japanese Self-Defence Forces (JSDF) in its review of its long-term threat assessment struck off Russia from the list and added China. Similarly, India has territorial disputes with China.

The two countries have fought a limited border war in 1962, in which India had to suffer a humiliating defeat at the hands of the Chinese forces. Any review of Indian literature pertaining to India's security and threat perceptions, instantly reveals that China looms large as the most likely adversary. India has not forgotten the humiliation in its war with China and considers it as unfinished business. Primarily it was the 'China threat' factor that forced India to sign a defence agreement with the former Soviet Union, despite its rhetorical non-aligned policy. In 1998, India again used the 'China threat' bogey to go ahead with its nuclear tests. More recently, the Indians perceive China's acquisition of naval facilities in Myanmar as a naval threat to what India interestingly considers as its own waters.

The Chinese help for developing port facilities at Bandar Abbas in Iran, Gawadar in Pakistan and naval facilities in Myanmar is viewed by the Indians with great anxiety. Perhaps that is one of the reasons that India has quickly moved to consolidate its relations with Iran and decided to open up a Consulate in Bandar Abbas. There is an evident and inherent clash of interests between China and India. It is this common concern over China's military growth, supported by a strong economy that provides a common ground for security cooperation between Japan and India.

The two countries have undertaken aggressive diplomatic initiatives to win over medium and small states in Asia, to neutralize the Chinese influence, by way of engagement. Indian engagement with Myanmar and Iran, the two countries which have defence cooperation with China, is a case in point. During the former Japanese Prime Minister Mori's visit to India in August 2000, he stated that, looking to the 21st century, Japan wanted to build a multifaceted cooperative relationship with India in a wide range of fields. He termed this relationship as the 'Global partnership between Japan and India in the 21st Century'.

The two sides expressed a desire to institutionalize a dialogue between the ministries of defence and foreign affairs for coordinated actions on security and foreign policy related issues, like the security of sea-lanes, joint naval exercises to combat piracy and disaster management etc. This outlined the level of Indo-Japanese security engagement. There are some common features that support a mutually cooperative framework:

    a. Both the countries are mature democracies and share common values.

    b. Both are seeking a global great power status.

    c. Both the countries are contenders for a permanent seat in the UN Security Council, in order to have a greater say in the global affairs. In this case the two countries have a desire to support each other's candidature for the UNSC seat in the capacity of developed and developing countries.

    d. Both the countries want to step up their involvement in Asian Affairs.

    e. Both the countries are members of ASEAN and ARF.

    f. Both the countries are dependent on petroleum imports from the Gulf and Middle

East.

g. Both have a stake in an un-interrupted supply of energy resources and safety of the Sea Lanes of Communications (SLOC).

h. Both the countries are extending their naval reach into the Malacca Straits and South China Sea as a counter to China's growing influence there.

It is primarily the last two factors, which have provided a basis for the cooperative security framework that entails amongst other policies joint naval exercises and joint patrolling of the SLOC, primarily against piracy.

SECURITY OF SLOC

In November 1999, the seizure of a Japanese ship, MV Alondra Rainbow, by pirates and its eventual recovery by the Indian Navy under their coordinated networking with international maritime agencies, highlighted the problems of piracy near the Straits of Malacca, which is one of the busiest and most important routes of sea trade.

The recovery of the ship supported the idea of joint patrolling in the region to deal effectively with such incidences. Japan and other ASEAN countries have a high stake in the safety of the lanes of communication. India has a significant naval build up at the Andaman and Nicobar islands, and has created a special Far Eastern Naval Command (FENC) based on these Islands.

THE STRATEGIC DIMENSION

Although the spelt out security cooperation framework between India-Japan and ASEAN countries is for the safety of SLOC and to meet the other non-traditional security threats, but the unstated contents of Indian involvement in the region suggest a deeper strategic dimension to it. As a security strategist observed, 'Naval diplomacy along with the new naval doctrine would form a critical component of India's policy towards the region'. In April 2000, India announced that it would hold bilateral naval exercises with both Vietnam and South Korea and also its plans to hold a unilateral exercise in the South China Sea.

The bilateral exercises were a manifestation of a more strategic 'Look East' policy. In 2000, the Indian Navy had sent warships, tankers and submarines to Japan, South Korea, Indonesia and Vietnam for bilateral exercises and as gestures of good will. As observed by an Indian political commentator:

'The holding of unilateral exercises in the South China Sea adds a different dimension. The South China Sea is the location of maritime disputes between several ASEAN countries and China over the Spratly Islands. According to a 1992 Chinese law, the South China Sea is considered to be Chinese territory. In these light, unilateral exercises by the Indian Navy involving several warships, a submarine and a maritime reconnaissance aircraft hope to assert India's prowess and establish freedom of navigation of that sea that only China considers its territory.

India hopes to send a signal to its South East Asian neighbours that it can challenge China's claims and begin to project itself as counterweight to China.' Countries in Asia have noticed with immense interest and a growing concern India developing its blue water navy and seeking opportunities for the projection of its naval power. China has watched with a degree of concern and annoyance, India's unilateral and joint naval exercises with countries like Vietnam, in the South China Sea. The Executive Director of the Australia Defence Association, Michael O' Connor, while commenting on the subject holds the view that,

'As a major military power, India has consciously moved its strategic focus eastwards into a region where it has no obvious security interests. India is moving out of the Indian Ocean, traditionally its area of blue water operations, into the Pacific through the Malaysia/Indonesia barrier. Even maintaining a base in South China Sea, with or without a base in Vietnam, impacts heavily upon South

East Asian security, so that Indonesia and Malaysia as well as Thailand will be forced to reconsider their relations not only with India and China but also with the U.S..

The emergence of India as a strategic player in South East Asia offers ASEAN the possibility of significant military support to replace that of Australia, backed as it is by the U.S..' The Indian involvement in the region's security sphere, in collaboration with the countries apprehensive of Chinese influence, would serve two main purposes besides many others. One, it would help contain China, which India can not do alone but very much desires to do so. Two, it would bring India closer to the U.S. and its NATO partners, besides creating dependence of South East Asian states in security terms on India.

INDO-VIETNAMESE SECURITY RELATIONS

India also forged new defence ties with its traditional Cold War friend Vietnam. India and Vietnam have already signed a defence protocol in 1994. In March 2000, during the five days visit by former Indian Defence Minister, George Fernandes, the framework of that protocol was expanded and a new defence agreement was signed. Under the new 15 point agreement, the salient features are as follows:

    a. India will repair and overhaul the Russian MiG aircraft fleet of the Vietnamese Air Force and also train its pilots.

    b. India will assist Vietnam in setting up of its defence industry and in manufacturing small and medium weapons and certain ordinance products.

    c. Vietnam to buy India's multi-role, advance light helicopter and fast patrol boats.

    d. India will provide expertise available with its Defence Research and Development Organization to assist Vietnamese victims of chemical warfare, specially of 'agent orange defoliant', excessively used by the Americans during the war.

e. The Indian Navy will help in the repair, up-gradation and building of vessels of the Vietnamese navy and train its technical personnel.

f. India will train Vietnamese officers in the application of information technology and software development in the defence field.

g. Vietnam will train Indian soldiers in jungle warfare.

h. The navies of India and Vietnam and the Indian Coast Guards and the Vietnamese sea-police to cooperate with each other in combating piracy.

Former Indian Defence Minister Fernandes, while speaking to the journalists during his March 2000, vist to Vietnam said, 'we have sought to raise defence contacts between the two countries to a much higher and larger plane.' The new defence pact paves the way for visits by Indian naval ships to Vietnamese ports and more collaboration in naval exercises.

At the same time, some sources have also reported that as India is inducting more sophisticated weaponry into its armed forces, it intends to sell its cast-off weapon systems at a very low price to Vietnam. The two countries' armed forces are overwhelmingly Russian-supplied.

Vietnam historically had adverse relations with China. It fought its last war with China in 1979. Though Vietnam has been able to peacefully resolve its land border dispute with China, the conflicting claims over the South China Sea remain to be resolved.

The mutual confidence between the two countries has not been fully restored. Presently, India and the U.S. are primarily considering Vietnam's importance to their anti-China stands, due to Vietnam's anti-China past and its geo-strategic location, in terms of its proximity to the SLOC area and to China. Both the U.S. and Indians are focusing their eyes on the Camaranh Bay, a naval base leased out to Russia. In 2004, the lease agreement of the naval base is to expire.

The U.S. is interested in that base to find an alternative for itself to replace its loss of Subic Bay in the Philippines. If this base facility is provided to India, it would enable India to maintain a permanent presence in the South China Sea. This would directly confront China's claims to sovereignty in that region. According to Nayan Chanda of the Far Eastern Economic Review, 'there are signs that an informal security cooperation chain is forming between Japan, Vietnam and India, all of whom share a common strategic concern in China'.

Former Indian Defence Minister, George Fernandes, had also hinted at a role for India in the South China Sea controversy, where the Philippines and Vietnam are challenging China's claims on the Spratly islands. Fernandes said, 'A strong India, economically and militarily well-endowed, will be a very solid agent to see that the sea lanes are not disturbed and that conflict situations are contained.' This statement clearly underscores the points mentioned earlier from the report of the Indian Ministry of Defence on Indian national security interests.

The U.S. apparently has approved India's new-sought role as a balancer against China's expanding influence in the Asia-Pacific region, along with U.S. traditional military allies like South Korea, Japan, Australia and Singapore. Professor Tomoda Seki of Asia University summed it up as, 'a triangulation with India in the west and Japan in the east of China could form the basis to keep any Chinese belligerence in check and the region at peace.'

India's entry into the regional 'balance of power' game may secure the safety of SLOC and help reduce non-traditional security threats. However, there are apprehensions expressed by experts on the region that India's entry into the region and its proclaimed intensions might restart a Cold War in Asia. Since India does not have direct vital security interests in the region, it can afford to raise the level of tensions in the region, because that serves well the emerging Indian arms industry and raising the costs for its main rival China in economic and political terms.

4

India's Strategy In The Indian Ocean

When India became independent from colonial rule in 1947, it chooses not to align with either of the East-West power blocs that were then taking shape. It did decide, however, to become a member of the British Commonwealth. At that time, Britain had a strategic concept for the defence of ˌthe Commonwealth against communism. In pursuance of that concept, the navies of India and other Commonwealth countries were offered reconditioned Second World War warships from Britain's reserve fleet, vessels that were surplus to British requirements. It was dear that the only way to remedy swiftly the after-effects of the division of the pre-partition navy between India and Pakistan was to continue the British connection and obtain whatever was offered and affordable.

India acquired a cruiser, some destroyers, and several smaller ships. Over the next few years, India placed orders in Britain for eight new frigates and initiated steps for the creation of a naval air arm and a submarine arm. It also decided to resume construction of warships, starting with frigates. Indian warship-building expertise had languished over the century since the transition from wooden to steel hulls.

By 1962, eight new frigates (mostly antisubmarine), a reconditioned aircraft carrier, and a second cruiser had arrived. Evaluations were still in progress regarding the frigate to be built in India (with European collaboration).

There had been no progress on the submarine arm; antisubmarine exercises were being seriously constrained by a lack of submarines with which surface ships could exercise. At the same time, a boundary dispute with China erupted into hostilities on the northern mountain borders. Indian ground forces suffered serious reverses. The United States responded positively to India's request for urgent military assistance. Pakistan, being an ally of the United States, felt discomfited and, acting on the dictum that "your enemy's enemy can be your friend," sought closer relations with China and to a lesser extent with the Soviet Union, the two countries that the Central Treaty Organization and Southeast Asia Treaty Organization were meant to contain.

China responded positively, initiating thereby the Pakistan-China geostrategic alignment in the Indian subcontinent. The postmortem on the military reverses of 1962 led to the formulation of India's first five-year defence plan. Its basic features were the immediate augmentation of the Army and the Air Force. The Navy, which had played no significant role in the conflict, was to continue its programme of replacing its old ships with newer ones. The Army, entrusted with the defence of the Andaman and Nicobar Islands since 1945, when Japan evacuated them, was relieved of that duty in 1962 by the Navy, to enable the Army to focus on the borders with China.

Britain agreed to train a few crews to man a submarine, so as to provide antisubmarine training. During 1964 defence delegations visited the United States, the Soviet Union, and Great Britain to explore ways of meeting the immediate requirements of India's defence plan. As regards the Navy, the U.S. response was to refer India to its traditional supplier, the United Kingdom. Britain, in turn, regretted that since it was pruning its own navy, it would not be able to meet India's requirements either for the latest types of destroyers and submarines that the Indian navy wanted or to extend financial support to build in Britain the modern submarines to start India's submarine arm.

An agreement was, however, signed for the construction in India, with British collaboration, of two British-designed Leander-class frigates. The Soviet Union, in contrast, offered to give the Indian Navy whatever it sought. Meanwhile, the regional maritime threat was increasing. In 1964, pursuant to the acquisition from the Russia of a large fleet from 1958 onward, there was a sharp rise in Indonesian bellicosity, and intrusions by that nation in the Andaman and Nicobar Islands increased. In 1965 hostilities erupted with Pakistan on two occasions. In the spring of 1965, Pakistani tanks (received from the United States as part of its military assistance programme) intruded into Indian Territory in the Rann of Kutch.

The memoirs of senior Pakistani officers reveal that the deployment of American-supplied armour in Kutch had two objectives. The first was to entice Indian armour away from northern India, where an attack on Kashmir was planned for later in the year, and the second was to see how strongly the United States would protest Pakistan's use of tanks it had provided, in clear violation of Pakistan's commitment. The United States did protest, but it was ignored. The second attack commenced in August. Intruders from Pakistan infiltrated Kashmir to sabotage vital installations, in the expectation of a spontaneous uprising by the local people. There was no uprising. The intruders were apprehended and the plan was revealed.

The Indian Army controlled the situation, and Pakistani morale collapsed. To restore spirits, the Pakistani Army itself crossed the international border into Kashmir on I September. The Indian Air Force halted the Pakistani tank columns despite fierce battles overhead between the two air forces. Pursuant to India's clear warning to Pakistan, given years earlier and often repeated thereafter, that "crossing the international border would invite strong retaliation," the Indian Army launched a counterattack on

6 September and advanced tc ward Lahore, in the Punjab. In response, the Paki tani land forces withdrew from Kashmir and headed for the Punjab. Land and air battles continued until a cease-fire was declared on 23 September.

The Indian fleet had been deployed in the east, in the Bay of Bengal, in August to deter any Indonesian naval intrusions in support of Pakistan. On I September, when the Pakistani Army crossed into Kashmir, the Indian fleet was ordered west to Mumbai (formerly Bombay), in the Arabian Sea. The fleet's ships were of varying vintage and had disparate speeds; they arrived in Mumbai in ones and twos from 7 September onward. Meanwhile, in reaction to the Indian Army's thrust into the Punjab, on the night of 7-8 December Pakistan sent its flotilla to carry out a bombardment of the coastal temple town of Dwarka, about two hundred miles south of the main naval base at Karachi, then return to its patrol area off Karachi, where it remained for the rest of the war.

When the Indian fleet had refueled and re-provisioned at Mumbai, it sailed to a patrol area off Saurashtra to deter further intrusions. Except for a large number of attacks against underwater contacts suspected to be the submarine that the United States had given Pakistan in 1964, no encounter occurred before the cease-fire. After the cease-fire there was considerable unhappiness within the Indian Navy. It had made no meaningful contribution to the war, and it had been unable to avenge the bombardment of Dwarka. Only later did it become generally known that the Indian government had directed the Navy to take no aggressive action at sea; the government had wanted to confine the scope of the fighting to land and air operations.

It also became known that Indonesia had dispatched a Russian-built submarine and some missile boats to assist the Pakistani Navy, though by the time they arrived the cease-fire had been declared. The events of 1965 indicated that the Navy would have to plan for concurrent operations in the Bay of Bengal in the east and the Arabian Sea in the west. This assessment, which coincided with still-pending plans of preceding years, precipitated several decisions. First, a new fleet would have to be created for the Bay of Bengal. This Eastern Fleet would have to be supported by a new dockyard and new logistic depots on the east coast of India.

The naval presence in the Andaman and Nicobar Islands would have to be increased, and maintenance facilities created so that patrol vessels would not have to undertake the long passage to the mainland. Further, India decided to accept the pending offer of the Russia to meet the Indian Navy's requirements for the latest ships and submarines; the new units would be based in the Bay of Bengal to counter Indonesian adventurism. Finally, in order to deter attacks on coastal ports, like that on Dwarka, Soviet missile boats of the type that had been supplied to the Indonesian and Egyptian navies were to be evaluated.

Between 1966 and 1971, most of these decisions were implemented. Five submarine chasers, two landing ships, five patrol boats, four submarines, a submarine depot ship, a submarine rescue vessel, and eight missile boats were acquired from Russia. Construction commenced of a new dockyard in Vishakhapatnam, where all Soviet-supplied vessels would be based, maintained, and refitted; all ships of Western origin were to be based at Mumbai. This arrangement was necessary to meet the Cold War concerns of the Russia regarding the leakage of its technology to the West, and also that of Britain (which had licensed the construction of Leander-class frigates in Mumbai) regarding the same to the East.

In March 1971, political ferment in East Pakistan (East Bengal) erupted into a struggle for secession from West Pakistan. Pakistan imposed martial law and ruthlessly suppressed the uprising. A subsequent commission headed by a judge of the Pakistan Supreme Court found that the Pakistani Army had resorted to genocide in an attempt to obliterate the aspirations of the people of East Bengal for independence. The major impact on India of this "internal affair" of Pakistan was a flood of refugees. Within a matter of months, over nine million Bengalis were living in refugee camps in India.

Infuriated East Bengalis, burning to avenge the brutalities they had suffered and the destruction of their homes, began guerrilla activity against Pakistan. For India, the situation became extremely difficult. The demographic composition of Indian border districts was changing to an extent that was politically unacceptable. Hawkish elements in India began calling for military action to stop the genocide and create conditions under which the refugees could go back. The Army was unprepared for military operations in the east. Appeals to the international community yielded generous humanitarian aid for the refugees but no answer to the problem of how and when the refugees could be made to feel safe enough to return to their homes in East Pakistan.

The Indian armed forces anticipated a Pakistani intrusion into India to eliminate the camps from which the guerrillas operated; the Army prepared to counter such an attack. The problem was complicated by Pakistan's declared strategy that "the defence of East Pakistan lay in the west," meaning that an attack by West Pakistan on India's western border would relieve Indian military pressure in the east. The situation was compounded by the likelihood of China's aiding Pakistan by forcing India to position troops to counter a Chinese threat on India's northeast frontier, where hostilities had occurred earlier in 1962, thereby forcing India to plan for hostilities on three fronts-the west, the east, and the northeast.

This geostrategic contingency was offset in August when India and the Soviet Union signed a twenty-year treaty of friendship. Between August and November there were several false alarms, but on 3 December 1971 the Pakistani Air Force attacked Indian airfields on the western border and initiated the war. Naval operations had an important role in the fourteen days of fighting that ensued; they marked the beginning of India's regional maritime eminence. In the Bay of Bengal, the Indian Navy's aircraft carrier and frigates enforced contraband control and choked off all resupply from seaward. Pakistan's U.S.-supplied submarine, which had been deployed in the east to seek and sink the Indian aircraft carrier, exploded and sank near the entrance to Vishakhapatnam harbour whilst trying to avoid an Indian warship.

The United States became apprehensive that should Pakistan's armed forces in the east collapse, India would transfer its forces from there to attack West Pakistan, which was an ally in the Central Treaty Organization. As a gesture of solidarity, on 10 September 1971 an American task force headed by the nuclear-powered aircraft carrier Enterprise was despatched from the Gulf of Tonkin toward the Bay of Bengal. On 6 and 13 December, the Soviet Navy despatched two groups of nuclear-missile-armed ships from Vladivostok; they trailed U.S. Task Force 74 in the Indian Ocean from 18 December until 7 January 1972. During the war, the Indian missile boats in the Arabian Sea had been divided into two groups.

One was deployed on the Saurashtra coast to attack ships off Karachi and to deter hit-and-run raids like the one that had occurred in 1965. The second group was assigned to the task force deployed in the Arabian Sea to enforce contraband control and attack Karachi from the southwest. The first Indian missile boat attack occurred on 4-5 December, from the south; it sank a destroyer and a coastal minesweeper. As a precaution, the Pakistani flotilla withdrew inside Karachi Harbour on 7 December. The second missile boat attack, which was made on 8-9 December from the southwest, hit the Pakistani Navy's tanker in the anchorage outside Karachi and set the oil storage tanks of Karachi on fire. Shipping traffic to and from Karachi ceased.

The Indian submarines were deployed off Pakistan's coast but did not encounter any warship targets. On 9

December a Pakistani submarine of the French Daphne class deployed off Saurashtra sank one of the two Indian antisubmarine frigates that had been despatched to nudge it to seaward and safeguard the forces assembling for the next missile attack. After the Pakistani forces in the east surrendered on 16 December, India offered Pakistan a cease-fire in the west, which Pakistan accepted on 17 December.

East Pakistan became the independent state of Bangladesh, and millions of Bengali refugees returned from India in early 1972. Fascinating vignettes of the complex geostrategic factors at work during this war can be found in the memoirs of President Richard Nixon, Dr. Henry Kissinger (his security advisor), Anatoly Dobrynin (then Soviet ambassador in Washington), and Admiral Elmo Zumwalt (then Chief of Naval Operations), and in the newspaper columns of Jack Anderson regarding the deliberations of the American government's decision-making body known as the Washington Special Action Group.

THE INDIAN NAVY'S DEVELOPMENT SINCE 1971

The Navy learned several lessons during the 1971 war that have governed its development in the thirty-one years since. The first was the need to maximise antisubmarine capability. The Navy has now acquired long, medium, and short-range antisubmarine aircraft, antisubmarine helicopters, hunter-killer submarines (from Germany), and, from diverse sources, longer-range sonars, torpedoes, and antisubmarine rockets for surface ships. A second lesson was the importance of defences against missiles fired from land, submarines, ships, and aircraft, for which several measures were necessary.

There had to be at least one more aircraft carrier, with aircraft capable of attacking missile-carrying platforms before they could launch their missiles. In addition, warships required electronic warfare equipment, antimissile missiles, and high-rate-of-fire guns for point defence. Third, older ships and submarines had to be replaced-by indigenous construction to the maximum extent possible, but in the meantime from abroad. A number were obtained from the Soviet Union, beginning in 1976, including Kashin-class destroyers, Nanuchka missile boats, minesweepers, and a tanker.

Destroyers, frigates, corvettes, and missile boats were built indigenously but with Russian weapons; domestically built ships without Soviet systems included amphibious vessels, a fleet tanker, offshore patrol vessels, survey ships, and patrol craft. In the 1980s the Navy acquired from Britain a secondhand aircraft carrier and vertical-takeoff-and-landing aircraft. All these surface ships replaced predecessors with in-service lives of about fifteen years for minor vessels and twenty-five years for principal warships. Plans for a more modern aircraft carrier are still under examination.

As regards submarines, Soviet Kilo-class boats replaced the Foxtrots. Four German conventional hunter-killer types were acquired, two built in Germany and two in India; plans are presently in hand to resume submarine construction in India. Fourth, it was learned that refit and repair facilities had to be augmented and kept in step with the latest equipment fitted in ships, submarines, and aircraft. The final lesson was that the Western and Eastern Fleets had to be kept trained for a modest but straightforward role-to deter aggression from seaward by posing a threat of punitive damage.

The three decades since the 1971 war have seen the development of the Indian Navy. Some have felt that the growth of the Navy has been slow, stunted by a lack of funds (because of preoccupation with the Army and Air Force) and by a lack of political and bureaucratic interest in maritime matters. Such views, however, are not borne out by the facts. Whilst this may have been said of particular five-year plans and provided grist for animated debate in professional circles, it is not tenable in a longer perspective. The Navy's growth has indeed been slow, but primarily as a necessary result of the long-term effects of certain decisions taken on major issues.

An example is the resolve to become self-reliant and constantly innovative-it takes years to develop the expertise and capacity needed for building the wide range of equipment that goes into modern ships and submarines. The Navy is also determined to procure the best that is available worldwide, integrating it with whatever equipment can be developed locally, and installing it in customised indigenous hulls. Similarly, a conscious choice has been made to forgo series production of major warships in favor of continuous improvement to technological capability, despite the penalties with respect to time, cost, and nonstandardisation.

Further, weapon and ship production is to be accompanied by the timely creation of modern facilities and depots to maintain a small but technologically contemporary navy. Lastly, the equipment suites of the Indian Coast Guard and the Navy are being harmonised to minimise, wherever possible, the Navy's coastal responsibilities during war. For a developing navy, such far-reaching decisions are noteworthy, considering the budgets that were available in the last few decades. That the Navy has been able to adhere to these plans is all the more remarkable in that it has had to survive the rigorous financial scrutiny that is characteristic of democratic governance.

During the Cold War, it was widely accepted that India would become embroiled in no confrontations except as part of United Nations peacekeeping operations, of which the Navy's deployment to Somalia was an example. In those years, ships of all navies happily visited Indian ports, and Indian ships showed the flag in other ports of the world. Except for the tasks of transporting the Army to Sri Lanka and back, and helping to snuff out the attempted coup in the Maldives (both operations having been carried out at the invitation of the respective governments), India and its navy seldom appeared on the strategic radar screens of the West.

In the years since the end of the Cold War, and particularly after India's nuclear tests in 1998, there has been an increased interest in both India and its naval capabilities, as can be seen in Western writings:

India now has neither an interest in challenging the system nor the means to do so, except marginally on nuclear issues, but it remains determined not to permit others to foreclose the possibility that it too may some day aspire to great-power status.

The strategies environment of Asia is characterized by the presence of three great continental powers - China, India, and Russia. Neither China nor India will have a true blue-water navy over the next five years—although they will both seek to extend their naval influence, and therefore their strategic ambitions will overlap in Southeast Asia. Whether Asia remains a peaceful region will largely depend upon the struggle for power and influence between the major powers: China, Japan, India, Russia, and the United States.

It is not in the interests of the United States or of its allies to see the region dominated by any one Asian power or by a concert of them.... As China's influence in Asia grows, India-which wants to be accepted as a major power- will seek to compete with China. Until recently, India's poor economic performance, its preoccupation with Pakistan, and earlier its alliance with the former Russia served to limit its interest elsewhere in Asia. But the Indian economy now seems to be set on a path of reform and is growing strongly. The military balance on the subcontinent now firmly favors India, and with each year that passes its superior economic performance will improve its military advantage. India, therefore, will be able to lift its strategic horizons.

Chinese policy is no longer driven by a felt need to counter reactively the growth of Indian power. Well prior to the collapse of the Soviet Union, Beijing and New Delhi were already exploring the modalities of a more stable relationship... The larger challenge for Beijing will be to pursue a more fully developed concept of future Sino-Indian relations that acknowledges India's primary in the regional balance of power, while still providing Pakistan the wherewithal to maintain autonomy from a presumptive Indian sphere of influence.

It will be many decades before India offers us bases, if it ever does. A number of "rationales" for a closer relationship with India:

1. India is a strategic counterweight to China.
2. India is the more "moderate," or "reachable one" regarding the burgeoning nuclear standoff with Pakistan.
3. India is a democracy in a region that has few others.
4. India is taking a distant backseat to China in attracting FDI [foreign direct investment] and U.S. government attention, thus precipitating behaviors designed to get Washington to "notice it" more.
5. India's naval buildup signals that it can play a serious stabilising or destabilising role in the all-important maritime sea lines of communication between the Middle East and Southeast Asia.
6. India is the obvious kingpin power in South Asia.
7. India, like China, is too big to ignore; but, unlike China, there is no sense of an emergent peer-competitor relationship.
8. India, like the U.S., is a former British colony, so there are good historical reasons for closer ties.

9. India's burgeoning role as a computer powerhouse in the global IT [information technology] economy, and the surprisingly large role of Indian expatriates in the U.S. IT sector, both inevitably lead to greater influence for India and Indian-Americans in U.S. foreign policy decision making.

The above perceptions are reasonable assessments of possibilities. Several other constructs could be equally reasonable. What would perhaps be especially helpful is to conclude with an Indian point of view.

THE INDIAN VIEW

India's achievements since independence, such as they are, are the products of two groups of factors. The first includes the sympathetic understanding of India's formidable developmental problems, as well as generous financial and technical assistance extended by the United States, the Soviet Union, Europe, Japan, and the oil-producing countries. The second comprises the ingenuity, innovativeness, and capacity for hard work that are so characteristic of the Indian people in finding solutions appropriate to Indian conditions.

If India is seen today as a country that is politically, economically, and militarily strong, it should also be remembered that in these fifty years or so, India has invariably exercised exemplary restraint in times of crisis. With this background, a number of realities about India may help to provide a framework for viewing the nation and its actions in the years ahead. India has never had, and does not now have, overseas territories or global national security interests requiring military capabilities. Nonetheless, India is a vast, well endowed subcontinent with sufficient indigenous resources to sustain its population at a tolerable level of welfare.

Inevitably-and India is very conscious of the fact-its size, economic strength, strategic depth, and population cause smaller neighbours to look upon it as a hegemon. Accordingly, India is always cautious to ensure that no action can be misinterpreted by hypersensitive neighbours as hegemonistic. It also realises that building mutual confidence takes decades. Despite four unsought wars and prolonged spells of bloody terrorism, India firmly believes that the only way to settle disputes is bilaterally across the negotiating table, however long it may take. The observation that "you have no idea how much it contributes to the general politeness and pleasantness of diplomacy when you have a little quiet armed force in the background" finds echoes in the Arthashastra, a classic Indian treatise on statecraft written in the third century B.C.

Nonetheless, in the field of global politics, India has steadfastly met all its financial, peacekeeping, and developmental commitments to the United Nations. Indian peacekeeping contingents have received universal praise from the time of the truce in Korea in 1952 to their present deployment in Africa. India has supported from the outset the United Nations resolution of 1971 that the Indian Ocean be a zone of peace. Today, thirty-one years later, when so much of the world's oil supplies are transiting the Indian Ocean, it is even more in the common interest that this ocean remains peaceful and that its sea-lanes remain free of tension.

For its part, India does not see that ocean as an "Indian Lake" and has never used this expression. Finally, there is no fundamental dash of interest between India and the United States, regionally or globally. Both share a heritage of being large, multiethnic, and democratic countries. Both share a particular interest in ensuring free and unthreatened navigation in the sea-lanes that carry oil to India and the rest of the world. India's draft nuclear doctrine has been officially opened to public debate. Its main elements are "no first use" and credible retaliatory capability. As and when the doctrine is finalized, the Indian Navy will prepare to provide the seaborne component of retaliation.

After roughly a decade in the strategic wilderness, the Indian Ocean region is again becoming an arena of geopolitical rivalry among world powers and local states. During the final decades of the Cold War, the region was a zone of fairly intense superpower competition. The United States and the Russia vied for political advantage, while their navies competed for refueling facilities and bases in places such as Socotra Island in the former South Yemen, Gan in the Maldives, and Port Victoria in the Seychelles.

The Indian Ocean was also significant in the nuclear arms race as both navies operated ballistic missile submarines in the region. Due to this clash of superpowers and Indian Ocean states' perception of these powers as interlopers in their region, the UN General Assembly declared the Indian Ocean a zone of peace in 1971. One year later, the United Nations created the Ad Hoc Committee on the Indian Ocean to find ways to implement this declaration. To date, despite some 450 meetings of this committee, the contemplated Zone of Peace still does not exist.

Moreover, key Western members of the Committee withdrew in 1989, arguing that superpower rivalry in the Indian Ocean had diminished with the end of the Cold War, rendering a Zone of Peace purposeless. A 1997 statement by U.S. Secretary of State Madeleine Albright even called the committee an example of UN financial wastefulness that should be disbanded. The observations by U.S. and other Western delegates were on the mark to a certain degree immediately after the Cold War ended. It was during the early 1990s that observers hoped for the beginning of a "New World Order" characterized by less confrontation and competition among all states.

However, from the vantage point of 2002, the international system does not appear to have evolved in this manner, and to an increasing degree it is confrontation and rivalry, not conciliation, that characterize interstate relations in the Indian Ocean region. A variety of factors have contributed to the intensifying strategic rivalry in the Indian Ocean. These are not the same explanatory factors that prevailed during the 1970s and 1980s. Rather, today's imperatives are more powerful and less transient than those of the past. One of these is the continued and growing importance of oil, energy, and other vital resources. The key prize is the Persian Gulf, which is accessible only via Indian Ocean shipping routes.

Currently about 25 percent of all oil used by the United States passes through Indian Ocean sea lanes and the Persian Gulf region, and the United States also depends on the Indian Ocean for the shipment of about 50 different strategic materials, including tin, nickel, iron, lead, and copper. Burma and Bangladesh, which are rich in minerals, are also seen by other key states as important future energy sources. Michael Klare has argued that the world is witnessing a growing competition over access to vital economic assets. In Foreign Affairs he writes that, "Because an interruption in the supply of natural resources would portend severe economic consequences, the major importing countries now consider the protection of this flow a significant national concern."

A SECOND FACTOR IN THE GROWING STRATEGIC SALIENCE OF THE INDIAN

Ocean is the so-called "rise" of India, the most important of the coastal states. As this power evolves, New Delhi's interest in the affairs of the ocean will increase, as will the interest shown by other states that will wish either to check India or to ally with it. Beijing will be particularly motivated to counter New Delhi.

For China, a key consideration is the one-third of its Gross Domestic Product attributable to foreign trade and the importance of the Indian Ocean to that trade. All Chinese commerce with Europe and petroleum imports from the Middle East traverse the Indian Ocean. China will not let India have a free hand in these waters and will seek to establish a position that allows for protection of its sea lanes.

China is also concerned that India has a geostrategic advantage in the Indian Ocean and could try to compensate for continental power inadequacies in any border conflict by taking action against Chinese interests in the Indian Ocean. Moreover, Chinese security officials believe Beijing needs ocean access because, in the event of a confrontation with the United States or Japan in East Asia, a strong Chinese presence in the Indian Ocean would help secure one of China's backdoors. Chinese strength in the Indian Ocean would also make it more difficult to sever its lines of communications — an element in Beijing's so-called "southern strategy"

These Chinese concerns mirror one of the key motives of the United States in raising its profile in this region in recent years and in the future. For the United States, China's increased presence in the Indian Ocean provokes the normal set of concerns of any existing power as it reacts to the challenge posed by a rising state. Since the mid-1990s, the United States has been acting to contain China through a series of security initiatives and bilateral arrangements in East and Southeast Asia.

More recently, however, the United States has begun to strengthen its security posture in South Asia and the Indian Ocean to further constrain China and consolidate a security network there before Washington is challenged by Beijing. It is in this context that we can understand the recent strengthening of U.S.-Indian relations and the statement of U.S. Secretary of State Colin Powell that "India has the potential to help keep the peace in the vast Indian Ocean area and its periphery.

We need to work harder and more consisten tly to help them in this endeavor." Nuclear weapons competition also has increased the region's profile. Key developments include the Indian and Pakistani nuclear tests in 1998, continuing nuclear and missile developments by both New Delhi and

Islamabad, and Iran's ongoing attempts to develop nuclear weapons and missiles.

In addition, a number of states either are now using or likely will soon employ the Indian Ocean as a patrol zone for submarines or surface warships equipped with nuclear-armed cruise missiles. The admiral commanding India's navy, for example, recently reaffirmed India's intention—consistent with New Delhi's draft nuclear doctrine—to disburse his country's deterrent equally among the three branches of the military. A Pakistani naval spokesman has said his country has similar plans. Moreover, it is possible that Israel, reacting to nuclear weapons developments in Iran, is already deploying nuclear-armed submarines to conduct Indian Ocean patrols.

In coming years, the Indian Ocean will be a patrol zone for Indian, Pakistani, and, at some point, Chinese submarines and surface warships. The region's importance is further augmented because it is home to the world's greatest concentration of Muslims. Decades ago, this may not have been an important consideration. Today, however, Islamic civilization often finds itself at odds with Israel, other Western powers, and Hindu India, and it will be in the Indian Ocean region that this conflict frequently will play out.

This is what occurred when the United States intervened in Somalia, when U.S. embassies were the targets of terrorism in Kenya and Tanzania, when the USS Cole was attacked in Yemen, and when New Delhi established a new command in the Andaman and Nicobar Islands, prompted partly by concerns about Islamic extremism in Indonesia's Aceh province.

The Indian Ocean, thus, maybe seen—depending upon one's viewpoint—as the front line in either the struggle against terrorism or the West's continuing "crusade" to contain the world of Islam. A final factor elevating the Indian Ocean region's strategic importance is the promising outlook for increased commerce, investment, and cultural and political interaction between Europe and the Asia-Pacific region.

Ties of this sort are well developed between North America and Europe and between North America and Asia, but the third side of this triangle—between Europe and Asia-Pacific—is underdeveloped. However, as both of these regions gain political and economic importance, this state of affairs will change; to a large extent, it will be through the air and sea lanes of the Indian Ocean and its key coastal cities that this increased interaction will occur.

THE U.S. PRESENCE

The principal states contending for power and influence in the Indian Ocean region are the United States, India, and China. The growing U.S. strategic profile around the Indian Ocean has been characterized by a continuing enhancement of the U.S. military presence in the region, renewed security links with Pakistan, and a growing relationship with India. The United States began to build its existing Indian Ocean strategic infrastructure in the years following the 1991 Persian Gulf War, and has particularly expanded efforts since 1995 when the administration of President Bill Clinton shifted toward a more interventionist international policy.

The United States increased its military presence by establishing the Fifth Fleet to oversee military activities in the region, and ships from the Fleet executed missile strikes against targets in Afghanistan and Sudan in 1998. In late 2001 and early 2002, U.S. warships in the Indian Ocean played a major role in U.S. military operations against the Taliban and Al Qaeda in Afghanistan. Although currently the U.S. Fifth Fleet is based in Bahrain, the proliferation of land-based anti-ship missiles and small attack craft is making the Persian Gulf an increasingly dangerous place to base large ships.

The United States may eventually relocate the fleet to an Indian Ocean port such as Muscat, Oman, which would allow the Navy to control access to the area with less danger to its ships. Oman has long served as a logistics and intelligence center for U.S. operations in the Persian Gulf and Arabian Sea. The U.S. military currently uses at least three air bases in Oman as part of operations in Afghanistan. Elsewhere in the Indian Ocean, an agreement with Singapore has allowed the United States to use the new Change deepwater aircraft carrier pier since March 2001. U.S. aircraft carriers and large-scale formations can enter the harbor for repairs, allowing U.S. warships more freedom in the South China Sea.

The United States, backed by Britain and France, also has intensified naval patrols off Somalia's coast and reconnaissance aircraft now fly over the country regularly. Furthermore, Washington has been pursuing an expanded security relationship with Pakistan and India. In addition to providing economic assistance in the months since the September 11 terrorist attacks, the United States, through Pentagon statements, has been planning the establishment of permanent U.S. military bases in Pakistan.

A strong presence there could easily provide support to the U.S. fleet in the Arabian Sea. In the event that the United States removes its troops from Saudi Arabia, Pakistan presents a promising option for relocation. Additionally, U.S. ties with India began to improve during the final year of the Clinton administration when U.S. President Bill Clinton visited India and Pakistan and forged an agreement with Indian Prime Minister Atal Behari Vajpayee to "create a closer and qualitatively new relationship."

The administration of President George Bush has further increased ties by meeting with high-level Indian government officials and sending U.S. diplomats on trips to India. The outcome of these visits has been the resumption of U.S. military sales to India, a plan to strengthen ties between the two nations' intelligence agencies, the initiation of joint exercises, and the first meeting of the U.S.-India Defence Policy Group in over four years.

REGIONAL MANEUVERS

Apart from the United States, India has done the most to enhance its posture in the Indian Ocean. For New Delhi, Burma is one of the most important zones of emphasis. Relations between Burma and China have been close in the past decade, and India has been troubled by a variety of Chinese activities in Burma including road building, the construction of military facilities, strong military relations, and the probable presence of Chinese intelligence facilities in the Andaman Sea. Responding to these activities, India has countered in the past two years by increasing its diplomatic presence in Burma, importing natural gas, and assisting in road building.

Elsewhere in the region, India has sought to strengthen relations with Indonesia, a country with which it has historically had strained ties. To this end, New Delhi and Jakarta recently signed a defence cooperation agreement creating a new security relationship. The Indian navy also has begun conducting joint military exercises with its Indonesian counterpart in the Andaman Sea and the Bay of Bengal.

Of equal importance are New Delhi's strengthened military capabilities in the region, grounded in better trained military forces, more modem equipment, and improved facilities. New installations in southern India will handle the berthing and repair of the naval fleet, and their construction reflects the shift in the Indian navy's focus of operations away from a preoccupation with Pakistan.

New airfields will extend India's strategic reach in the Indian Ocean and Bay of Bengal and constitute a response to China's alleged interest in projecting power in this direction.

The emphasis on new bases has been paralleled by a marked and relatively successful effort by India in the past two years to acquire new and more lethal military equipment, including guided missile destroyers, refitted aircraft carriers, supersonic and-ship cruise missiles, and 1500-ton Scorpene submarines to be imported from France. India is in negotiations with Russia for nuclear-powered submarines and an additional aircraft carrier.

This list of developments is impressive and contrasts sharply with India's poor record of defence modernization during most of the 1980s and 1990s. Recently, China's profile in the Indian Ocean region has been growing less rapidly, though several initiatives are worth noting.

One is the November2001 agreement with the Association of Southeast Asian Nations to establish a free trade area within the next decade. This agreement would give China increased influence in this part of the Indian Ocean.

A second development was Beijing's dispatch, concurrent with New Delhi's recent threat to go to war against Pakistan, of five ships reportedly loaded with military equipment to Karachi. Pakistani President Pervez Musharraf visited China twice at the height of those tensions. Third, China announced on January 22,2002, that it is pledging US$4 billion to finance construction of a railroad between Kunming in Yunnan Province and Bangkok, Thailand, as part of the UN-sanctioned Trans-Asian Railway.

And finally, a new Chinese-built port and shipyard was inaugurated in Burma at Thilawa, 25 kilometers south of Rangoon, designed to cater primarily to Chinese ships. Beijing's ties with Pakistan have been its most important source of influence in the region. These have consisted of important military relations, continuing Chinese assistance to Pakistan's nuclear weapons program, and aid in building the new Pakistani port at Gwadar.

Notwithstanding these links, China's relative influence in Pakistan, while still quite significant, appears to have waned recently as U.S. clout has increased. The new U.S.

military presence at Pasni in Pakistani Balochistan is of special concern to Beijing, which has long hoped to establish a base along the Balochistan coast as a bulwark against the U.S. presence in the Persian Gulf. But no progress has been possible on these projects since the first U.S. troops arrived in the region after September 11. The other key element in China's Indian Ocean strategy has been Burma.

Here the most significant Chinese markers have been its military presence on Burma's coast and on nearby islands in the Andaman Sea, China's continuing influence over the Burmese military, and Beijing's hopes and plans for a security and commercial alternative to the Pacific Ocean— an "Irrawaddy corridor" linking China directly to the Indian Ocean through Burma.

Burma houses military compounds identified as intelligence facilities, radar stations, and Chinese satellite stations. During the past two years, Chinese navy and air force personnel have started helping Burma build new naval and air bases.

In the past decade, China also has supplied several billion U.S. dollars worth of arms to Burma in preparation for a feared Western or Indian intervention on behalf of Burma's democratic opposition. Beijing also has attached considerable importance to its relations with Burma because of the latter's potential as an avenue for commerce between China's southwest and both the Indian Ocean and Southeast Asia.

To this end, and through much of the 1990s, there were numerous reports of Chinese assistance, suggesting that an "Irrawaddy corridor" connecting China's Yunnan Province and the Bay of Bengal would be opening in the near future. Since mid-1997, however, apparently because of Burmese anxiety about potentially too-close relations with China, work on this corridor appears to have slowed.

Reacting to this trend, Chinese President Jiang Zemin conducted a four day visit to Burma in December 2001. During the trip the Chinese government promised US$100 million in Chinese investment to fund 12 projects focusing on agriculture, infrastructure, and human-resource development.

RISING PROMINENCE

The renewed rivalry in the Indian Ocean is not surprising, but it is troubling. It was anticipated by security observers in the 1970s and 1980s, and their concern led them to propose the still unrealized "Indian Ocean Zone of Peace." The competition for resources increases the likelihood of conflict between China and India and encourages more lethal forms of violence in a region where fighting was once unlikely. Even short of open conflict, the rivalry involving the United States, India, and China will force the smaller states of the region to choose sides and become increasingly entangled in a rivalry that will not serve their interests.

China, for example, can be expected to redouble its influence-seeking in Nepal, Bangladesh, Sri Lanka, and elsewhere if, in fact, it perceives that its strategic position in the Indian Ocean has been weakened by recent Indo-U.S. initiatives. One must also question whether the United States, by involving itself to an unprecedented degree in a region far from its shores, is acting wisely. In the past, the Indian Ocean region and East Asia have been distinct and largely autonomous security sub-regions.

Wars in Korea, Vietnam, and Kashmir have been limited geographically. Now, however, the new rivalry in the Indian Ocean is linking the East Asian security subsystem with that of South Asia, and future conflicts will become less geographically limited and more dangerous. One of the dominant features of global political geography in the 20th century was the prominence of northern Europe and northeastern Asia. It was in these areas that the great powers were concentrated, the great wars occurred, and the key international political struggles of the century unfolded.

We do not yet know what parts of the world will become important conflict zones in this century, but no region is as likely to play a crucial role as the Indian Ocean due to its combination of oil, Islam, and the likely rivalry between India and China. The Mediterranean was the center of strategic rivalry in times past. The Atlantic and the Pacific played this role in the 20th century. Based on recent developments, it is likely that the Indian Ocean region will surpass both of these zones in importance in the 21st century. The Indian Ocean, often characterized in the past as "the neglected ocean," will be so no longer.

5

China And U.S. Strategy In The Indian Ocean

The Asian seas today are witnessing an intriguing historical anomaly - the simultaneous rises of two homegrown maritime powers against the backdrop of U.S. dominion over the global commons. The drivers behind this apparent irregularity in the Asian regional order are, of course, China and India. Their aspirations for great-power status and, above all, their quests for energy security have compelled both Beijing and New Delhi to redirect their gazes from land to the seas. While Chinese and Indian maritime interests are a natural outgrowth of impressive economic growth and the attendant appetite for energy resources, their simultaneous entries into the nautical realm also portend worrisome trends.

PROSPECTS FOR A STRATEGIC TRIANGLE

At present, some strategists in both capitals speak and write in terms that anticipate rivalry with each other. Given that commercial shipping must traverse the same oceanic routes to reach Indian and Chinese ports, mutual fears persist that the bodies of water stretching from the Persian Gulf to the South China Sea could be held hostage in the event of crisis or conflict.1 Such insecurities similarly animated naval competition in the past when major powers depended on a common nautical space. Moreover, lingering questions over the sustainability of American primacy on the high seas have heightened concerns about the U.S. Navy's ability to guarantee maritime stability, a state of affairs that has long been taken for granted. It is within this more fluid context that the Indian Ocean has assumed greater prominence.

Unfortunately, much of the recent discourse has focused on future Chinese naval ambitions in the Indian Ocean and on potential U.S. responses to such a new presence. In other words, the novelty, as it currently stands, of the Indian Ocean stems from expected encounters between extra-regional powers.

But such a narrow analytical approach assumes that the region will remain an inanimate object perpetually vulnerable to outside manipulation. Also, more importantly, it overlooks the possible interactions arising from the intervention of India, the dominant regional power. Indeed, omitting the potential role that India might play in any capacity would risk misreading the future of the Indian Ocean region.

There is, therefore, an urgent need to bring India more completely into the picture as a full participant, if not a major arbiter, in the region's maritime future. In order to add depth to the existing literature, this chapter assesses the longer-term maritime trajectory of the Indian Ocean region by examining the triangular dynamics among the United States, China, and India. To be sure, the aspirational nature of Chinese and Indian nautical ambitions and capabilities at the moment precludes attempts at discerning potential outcomes or supplying concrete policy prescriptions.

Nevertheless, exploring the basic foundations for cooperation or competition among the three powers could provide hints at how Beijing, Washington, and New Delhi can actively preclude rivalry and promote collaboration in the Indian Ocean. As a first step in this endeavor, this chapter examines a key ingredient in the expected emergence of a "strategic triangle" - the prospects of Indian sea power.

While no one has rigorously defined this international-relations metaphor, scholars typically use it to convey a strategic interplay of interests among three nation-states. In this initial foray, we employ the term fairly loosely, using it to describe a pattern of cooperation and competition among the United States, China, and India.

It is our contention that Indian Ocean stability will hinge largely on how India manages its maritime rise. On the one hand, if a robust Indian maritime presence were to fail to materialize, New Delhi would essentially be forced to surrender its interests in regional waters, leaving a strategic vacuum to the United States and China. On the other hand, if powerful Indian naval forces were one day to be used for exclusionary purposes, the region would almost certainly become an arena for naval competition.

Either undesirable outcome would be shaped in part by how India views its own maritime prerogatives and by how Washington and Beijing weigh the probabilities of India's nautical success or failure in the Indian Ocean. If all three parties foresee a muscular Indian naval policy, then, a more martial environment in the Indian Ocean will likely take shape.

But if the three powers view India and each other with equanimity, the prospects for cooperation will brighten considerably. Capturing the perspectives of the three powers on India's maritime ambitions is thus a critical analytical starting point.

To provide a comprehensive overview of each capital's estimate of future Indian maritime power, this chapter gauges the current literature and forecasts in India, the United States, and China on Indian maritime strategy, doctrine, and capabilities. It then concludes with an analysis of how certain changes in the maritime geometry in the Indian Ocean might be conducive to either cooperation or competition.

INDIA'S SELF-ASSESSMENT

While Indian maritime strategists are not ardent followers of Alfred Thayer Mahan, they do use him to underscore the importance of the Indian Ocean. A Mahan quotation (albeit of doubtful provenance) commonly appears in official and academic discussions of Indian naval power, including the newly published Maritime Military Strategy. That is, as an official Indian press release declared in 2002, "Mahan, the renowned naval strategist and scholars, had said over a century ago, 'whosoever controls the Indian Ocean, dominates Asia. In the 21st century, the destiny of the world will be decided upon its waters.'"

Rear Admiral R. Chopra, then the head of sea training for the Indian Navy, offered a somewhat less bellicose-sounding but equally evocative version of the quotation at a seminar on maritime history: "Whoever controls the Indian Ocean controls Asia. This ocean is the key to the Seven Seas."

Quibbles over history aside, India clearly sees certain diplomatic, economic, and military interests at stake in Indian Ocean waters. In particular, shipments of Middle East oil, natural gas, and raw materials are crucial to India's effort to build up economic strength commensurate with the needs and geopolitical aspirations of the Indian people. Some 90 percent of world trade, measured by bulk, travels by sea. A sizable share of that total must traverse narrow seas in India's geographic neighborhood, notably the straits at Hormuz, Malacca, and Babel Mandeb. Shipping is at its most vulnerable in such confined waterways. Strategists in New Delhi couch their appraisals of India's maritime surroundings in intensely geopolitical terms - jarringly so for Westerners accustomed to the notion that economic globalization has rendered power politics and armed conflict passé.

The Indian economy has grown at a rapid clip - albeit not as rapidly as China's - allowing an increasingly confident Indian government to yoke hard power, measured in ships, aircraft, and weapons systems, to a foreign policy aimed at primacy in the Indian Ocean region. If intervention in regional disputes or the internal affairs of South Asian states is necessary, imply Indian leaders, India should do the intervening rather than allow outsiders any pretext for doing so. Any doctrine aimed at regional preeminence will have a strong seafaring component. In 2004, accordingly, New Delhi issued its first public analysis of the nation's oceanic environs and of how to cope with challenges there.

Straightforwardly titled Indian Maritime Doctrine, the document describes India's maritime strategy largely as a function of economic development and prosperity:

India's primary maritime interest is to assure national security. This is not restricted to just guarding the coastline and island territories, but also extends to safeguarding our interests in the [exclusive economic zone] as well as protecting our trade. This creates an environment that is conducive to rapid economic growth of the country. Since trade is the lifeblood of India, keeping our SLOCs [sea lines of communication] open in times of peace, tension or hostilities is a primary national maritime interest.

The trade conveyed by the sea-lanes traversing the Indian Ocean ranks first among the "strategic realities" that the framers of the Indian Maritime Doctrine discern. Roughly forty merchantmen pass through India's "waters of interest" every day. An estimated $200 billion worth of oil transits the Strait of Hormuz annually, while some $60 billion transits the Strait of Malacca en route to China, Japan, and other East Asian countries reliant on energy imports. India's geographic location and conformation rank next in New Delhi's hierarchy of strategic realities. Notes the Indian Maritime Doctrine, "India sits astride ... major commercial routes and energy lifelines" crisscrossing the Indian Ocean region.

Outlying Indian possessions such as the Andaman and Nicobar islands sit athwart the approaches to the Strait of Malacca, while the Persian Gulf is near India's western coastline, conferring a measure of influence over vital sea communications to and from what amounts to a bay in the Indian Ocean. While geography may not be destiny, the document states bluntly that "by virtue of our geography, we are ... in a position to greatly influence the movement/security of shipping along the SLOCs in the [Indian Ocean Region] provided we have the maritime power to do so. Control of the choke points could be useful as a bargaining chip in the international power game, where the currency of military power remains a stark reality".

The Indian Maritime Doctrine prophesies a depletion of world energy resources that will make the prospect of outside military involvement in India's geographic environs even more acute than it already is. The dependence of modern economies on the Gulf region and Central Asia "has already invited the presence of extra- regional powers and the accompanying Command, Control, Surveillance and Intelligence network. The security implications for us are all too obvious." Sizable deposits of other resources - uranium, tin, gold, diamonds - around the Indian Ocean littoral only accentuate the factors beckoning the attention of outside maritime powers to the region.

Indian leaders, then, take a somber view of the international security environment. In the "polycentric world order" New Delhi sees taking shape, economics is "the major determinant of a nation's power." While "India holds great promise," owing to its size, location, and economic acumen, its "emergence as an economic power will undoubtedly be resisted by the existing economic powers, leading to conflicts based on economic factors." The likelihood that competitors will "deny access to technology and other industrial inputs," combined with "the shift in global maritime focus from the Atlantic-Pacific combine to the Pacific-Indian Ocean region," will only heighten the attention major powers pay to the seas.

A buildup of Indian maritime power represents the only prudent response to strategic conditions that are at once promising and worrisome in economic terms. Maritime threats fall into two broad categories, in the Indians' reckoning. First, judging from official pronouncements such as the maritime doctrine and the newly published Maritime Military Strategy, New Delhi is acutely conscious that such nontraditional threats as seagoing terrorism, weapons proliferation, or piracy could disrupt vital sea-lanes. Cleansing Asian waters of these universal scourges has become a matter of real and growing concern.

Second, Indians are wary not only of banditry and unlawful trafficking but also of rival navies. While Indian strategists exude growing confidence, increasingly looking beyond perennial nemesis Pakistan, they remain mindful of the Pakistani naval challenge, a permanent feature of Indian Ocean strategic affairs. Over the longer term, a Chinese naval buildup in the Indian Ocean, perhaps centered on Beijing's much-discussed "string of pearls," would represent cause for concern.

This is the most likely quarter from which a threat to Indian maritime security could emanate over the long term, once China resolves the Taiwan question to its satisfaction and is free to redirect its attention to important interests in other regions - such as free passage for commercial shipping through the Indian Ocean region. But Indians remain acutely conscious that the U.S. Navy rules the waves in Asia, as it has since World War II. Despite closer maritime ties with the United States, Indian officials bridle at memories of the Seventh Fleet's intervention in the Bay of Bengal during the 1971 Indo-Pakistani war.

They also remain ambivalent about the American military presence on Diego Garcia, which they see as an American beachhead in the Indian Ocean region. Observes one Indian scholar, Diego Garcia and the Bengal naval deployment have "seeped into Indians' cultural memory - even among those who know nothing about the sea." Whatever the prospects for a U.S.-Indian strategic partnership, such memories will give rise to a measure of wariness in bilateral ties. On balance, the factors impinging on Indian and U.S. strategic calculations will make for some form of partnership-but perhaps not the grand alliance American leaders seem to assume. Even partnership is not a sure thing, however, and sustaining it will require painstaking work on both sides.

MODELS FOR INDIAN SEA POWER

The challenges it perceives as it surveys India's surroundings and the novelty of Indian pursuit of sea power have induced New Delhi to consult Western history. That Indians would look to American rather than European history for guidance, however, may come as a surprise. Given their skepticism toward American maritime supremacy - the residue of Cold War ideological competition, as well as a product of geopolitical calculations- nineteenth-century American history represents an unlikely source for lessons to inform the efforts of Indians to amass maritime power. There is a theoretical dimension to India's maritime turn as well.

Many scholars of "realist" leanings assume that the sort of balance-of-power politics practiced in nineteenth-century Europe will prevail in Asia as the rises of China and India reorder regional politics. If so, the coming years will see Asian statesmen jockeying for geopolitical advantage in the manner of a Bismarck or Talleyrand. There is merit to objections to the notion that strategic triangles and similar metaphors are artifacts of nineteenth-century thinking, and many Indians and Chinese think in geopolitical terms reminiscent of that age.

Other scholars deny that European-style realpolitik is universal, predicting instead a revival of Asia's hierarchical, China-centric past. Chinese diplomats have skillfully encouraged such notions, hinting that a maritime order presided over by a capable, benevolent China - and excluding predatory Western sea powers such as America -

would benefit all Asian peoples, now as in bygone centuries. Indians more commonly look for insight to a third model - the Monroe Doctrine, the nineteenth-century American policy declaration that purported to place the New World off limits to new European territorial acquisitions or to any extension of the European political system to American states not already under Europe's control.

James Monroe and John Quincy Adams (the architects of the Monroe Doctrine), Grover Cleveland and Richard Olney (who viewed the doctrine as a virtual warrant for U.S. rule of the Americas), and Theodore Roosevelt (who gave the doctrine a forceful twist of his own) may exercise as much influence in Asia - particularly South Asia - as any figure from European or Asian history. Soon after independence, Indian statesmen and pundits took to citing the Monroe Doctrine as a model for Indian foreign policy. It is not entirely clear why Indians adopted a Western paradigm for their pursuit of regional preeminence rather than some indigenous model suited to South Asian conditions.

India's tradition of nonalignment surely played some role in this, however. For one thing, Monroe and Adams announced their doctrine in an era when American nations were throwing off colonial rule, while India's security doctrine had its origins in the post-World War II era of decolonization. Thus the United States of Monroe's day, like newly independent India, positioned itself as the leader of a bloc of nations within a geographically circumscribed region, resisting undue political influence - or worse - from external great powers. This imparts some resonance to Monroe's principles despite the passage of time and the obvious dissimilarities between American and Indian histories and traditions.

Thus the diplomatic context was apt - especially since Indian statesmen intent on effective "strategic communications" designed their policy pronouncements to appeal to not only domestic but also Western audiences.

Prime Minister Jawaharlal Nehru's speech justifying the use of free to evict Portugal from the coastal enclave of Goa is worth quoting at length:

Even some time after the United States had established itself as a strong power, there was the fear of interference by European powers in the American continents, and this led to the famous declaration by President Monroe of the United States [that] any interference by a European country would be an interference with the American political system. The Portuguese retention of Goa is a continuing interference with the political system established in India today. Any interference by any other power would also be an interference with the political system of India today. ... It may be that we are weak and we cannot prevent that interference. But the fact is that any attempt by a foreign power to interfere in any way with India is a thing which India cannot tolerate, and which, subject to her strength, she will oppose. That is the broad doctrine I lay down.

Parsing Nehru's bracing words, the following themes emerge. First, while a European power's presence in South Asia precipitated his foreign-policy doctrine, he forbade any outside power to take any action in the region that New Delhi might construe as imperiling the Indian political system.

This was a sweeping injunction indeed. Second, he acknowledged the realities of power but seemingly contemplated enforcing his doctrine with new vigor as Indian power waxed, making new means and options available. Third, Nehru asked no one's permission to pursue such a doctrine. While this doctrine would not qualify as international law, then, it was a policy statement to which New Delhi would give effect as national means permitted. India did expel Portugal from Goa in 1961 - affixing an exclamation point to Nehru's words.

Prime Ministers Indira Gandhi and Rajiv Gandhi were especially assertive about enforcing India's security doctrine. From 1983 to 1990, for example, New Delhi applied political and military pressure in an effort to bring about an end to the Sri Lankan civil war. It c ployed Indian troops to the embattled island, waging a bitter counterinsurgent campaign - in large part because Indian leaders feared that the United States would involve itself in the dispute, in the process obtaining a new geostrategic foothold at Trincomalee, along India's southern flank.

One commentator in India Today interpreted New Delhi's politico-military efforts as "a repetition of the Monroe Doctrine, a forcible statement that any external forces prejudicial to India's interests cannot be allowed to swim in regional waters."

India's security doctrine also manifested itself in 1988, when Indian forces intervened in a coup in the Maldives, and in an 1989-90 trade dispute with Nepal. A Western scholar, Devin Hagerty, sums up Indian security doctrine thus:

The essence of this formulation is that India strongly opposes outside intervention in the domestic affairs of other South Asian nations, especially by outside powers whose goals are perceived to be inimical to Indian interests. Therefore, no South Asian government should ask for outside assistance from any country; rather, if a South Asian nation genuinely needs external assistance, it should seek it from India. A failure to do so will be considered anti-Indian.

This flurry of activity subsided after the Cold War, as the strategic environment appeared to improve and New Delhi embarked on an ambitious program of economic liberalization and reform. Even so, influential pundits - even those who dispute the notion of a consistent Indian security doctrine - continue to speak in these terms.

Indeed, they seemingly take the concept of an Indian Monroe Doctrine for granted. C. Raja Mohan, to name one leading pundit, routinely uses this terminology, matter-of-factly titling one op-ed column "Beyond India's Monroe Doctrine" and in another exclaiming that "China just tore up India's Monroe Doctrine."

Speaking at the U.S. Naval War College in November 2007, Rear Admiral Chopra vouchsafed that India should "emulate America's nineteenth century rise" to sea power. As India's naval capabilities mature, matching ambitious ends with vibrant means, its need to cooperate with outside sea powers will diminish. Declared Chopra, New Delhi might then see fit to enforce "its own Monroe Doctrine" in the region. The doctrine has entered into India's vocabulary of foreign relations and maritime strategy. Again, using nineteenth-century American history as a proxy, we can discern three possible maritime futures for India:

"Monroe." Indian statesmen animated by Monroe's principles as originally understood would take advantage of the maritime security furnished by a dominant navy, dedicating most of their nation's resources and energies to internal development. Limited efforts at suppressing piracy, terrorism, and weapons trafficking - the latter-day equivalents to the slave trade, a scourge the U.S. and Royal navies worked together to suppress - would be admissible under these principles, as would disaster relief and other humanitarian operations intended to amass goodwill and lay the groundwork for more assertive diplomatic ventures in the future. This modest reading of the Monroe Doctrine would not forbid informal cooperation with the U.S. Navy, today's equivalent to the Royal Navy of Monroe's day.

In 1895, President Grover Cleveland's secretary of state, Richard Olney, informed Great Britain that the American "fiat is law" throughout the Western Hemisphere, by virtue of not only American enlightenment but also physical might - the republic's capacity to make good on Monroe's precepts. This hypermuscular vision of the Monroe Doctrine would impel aspirants to sea power to avow openly their desire to dominate surrounding waters and littoral regions. From a geographic standpoint, the Cleveland/Olney model would urge them to make good on their claims to regional supremacy, employing naval forces to project power throughout vast areas.

No international dispute would be off limits that national leaders deemed a threat to their interests, and they would evince a standoffish attitude toward proposals for cooperation with external naval powers. Theodore Roosevelt took a preventive view of the Monroe Doctrine, framing "an international police power" that justified American intervention in the affairs of weak American states when it appeared that Europeans might use naval force to collect debts owed their lenders - and, in the process, wrest naval stations from states along sea-lanes vital to U.S. shipping.

TR's interpretation of the Monroe Doctrine, as expressed in his 1904 "corollary" to it, called for a defensive posture: Monroe's principles applied when vital national interests were at stake, and the would-be dominant power could advance its go od-government ideals. These principles would apply, however, within circumscribed regions of vital interest and be implemented with circumspection, using minimal force, and that in concert with other tools of national power. Cooperation with outside powers with no likely desire or capacity to infringe on the hegemon's interests would be acceptable.

What form such a doctrine will assume, and how vigorously New Delhi prosecutes it, will depend on such factors as Indian history and traditions, the natures and magnitudes of the security challenges Indians perceive in the Indian Ocean, the vagaries of domestic politics, and the Indian Navy's ability to make more than fitful progress toward fielding potent naval weapon systems. India will pursue its doctrine according to its needs and capabilities - just as each generation of Americans reinterpreted the Monroe Doctrine to suit its own needs and material power.

AMERICAN VIEWS OF INDIAN SEA POWER

Curiously, given the importance they attach to the burgeoning U.S.-Indian relationship and their concerted efforts to forge a seagoing partnership, American policy makers and maritime strategists have paid scant attention to the evolution of Indian sea power or the motives and aspirations prompting New Delhi's seaward turn. One small example: the Pentagon publishes no Indian counterpart to its annual report The Military Power of the People's Republic of China, despite the growth of Indian power and ambition.

To the contrary: American diplomats speak in glowing terms of a "natural strategic partnership" between "the world's biggest" and "the world's oldest" democracies, while the U.S. military has reached out to the Indian military on the tactical and operational levels - through, for example, the sixteen-year-old Malabar series of combined maritime exercises. Few in Washington have devoted much energy to what lies between high diplomacy and hands-on military-to-military cooperation, to analyzing the maritime component of Indian grand strategy.

True, the recently published U.S. Maritime Strategy, A Cooperative Strategy for 21st Century Seapower, proclaims that "credible combat power will be continuously postured in the Western Pacific and the Arabian Gulf/Indian Ocean," but its rationale for doing so is purely functional in nature: guarding American interests, assuring allies, deterring competitors, and so forth. The multinational context for this pronouncement - how Washington ought to manage relations with regional maritime powers, such as India, on which the success of a cooperative maritime strategy ineluctably depends - is left unexplained.

Why New Delhi has rebuffed such seemingly uncontroversial U.S.-led ventures as the Proliferation Security Initiative (PSI), a primarily maritime effort to combat the traffic in materiel related to weapons of mass destruction, and Task Force 150, the multinational naval squadron monitoring for terrorists fleeing Afghanistan, will remain a mystery to American officials absent this larger context. Why the apparent complacency toward India on the part of U.S. officials? Several possible explanations come to mind. For one thing, the United States does not see India as a threat.

The Clinton and Bush administrations have enlisted New Delhi in a "Concert of Democracies," and, as mentioned before, they view India as a natural strategic partner or ally. For another, other matters have dominated the bilateral relationship in recent years. The Bush administration lifted the sanctions imposed after the 1998 Indian and Pakistani nuclear tests and negotiated an agreement providing for transfers of American nuclear technology to the Indian commercial nuclear sector in exchange for partial international supervision of Indian nuclear facilities. Legislative approval of this "123" agreement remains uncertain, in large part because of questions as to whether new Indian nuclear tests would terminate the accord. Maritime cooperation has been subsumed in other issues.

Also, and more to the point, India has been slow to publish a maritime strategy that American analysts can study. Its Maritime Doctrine appeared in 2004, but a full-fledged maritime military strategy appeared only in 2007 - meaning that India watchers in the United States have had little time to parse its meaning and its implications for U.S.-Indian collaboration at sea, let alone to publish and debate their findings. For now, absent significant policy attention, any maritime-strategic partnership will take place on the functional level, with "naval diplomacy" filling the void left by policy makers.

How Washington will grapple with Indian skepticism toward the PSI and other enterprises remains to be seen. If New Delhi does indeed embark on a Monroe Doctrine - especially one of the more militant variants identified above - political supervision of U.S. naval diplomacy will be at a premium for Washington. Should the nuclear deal falter in Congress, for example, will that further affront the sensibilities of Indians intent on regional primacy? If so, with what impact on American mariners' efforts to negotiate a good working relationship at sea? The opportunity to craft a close strategic partnership with New Delhi could be a short-lived one as Indian power grows, especially if Indian leaders take an ominous view of their nation's geopolitical surroundings or if irritants to U.S.-Indian relations begin to accumulate.

CHINESE VIEWS OF INDIAN SEA POWER

If American analysts seem blasé about the intentions and capabilities of their prospective strategic partner, many Chinese analysts depict the basic motives behind India's maritime ambitions in starkly geopolitical terms. Indeed, their assumptions and arguments are unmistakably Mahanian. Zhang Ming of Modern Ships asserts, "The Indian subcontinent is akin to a massive triangle reaching into the heart of the Indian Ocean, benefiting any from there who seeks to control the Indian Ocean."

In an article casting suspicion on Indian naval intentions, the author states, "Geo-strategically speaking, the Indian Ocean is a link of communication and oil transportation between the Pacific and Atlantic Oceans and India is just like a giant and never-sinking aircraft carrier and the most important strategic point guarding the Indian Ocean." The reference to an unsinkable aircraft carrier was clearly meant to trigger an emotional reaction, given that for many Chinese the phrase is most closely associated with Taiwan. Intriguingly, some have invoked Mahanian language, wrongly attributed to Mahan himself, to describe the value of the Indian Ocean to New Delhi.

One Chinese commentator quotes (without citation) Mahan as asserting, "Whoever controls the Indian Ocean will dominate India and the coastal states of the Indian Ocean as well as control the massive area between the Mediterranean and the Pacific Ocean." In a more expansive reformulation, two articles cite Mahan as declaring, "Whoever controls the Indian Ocean controls Asia. The Indian Ocean is the gateway to the world's seven seas. The destiny of the world in the 21st century will be determined by the Indian Ocean." (As noted before, a very similar, and likewise apocryphal, Mahan quotation has made the rounds in India - even finding its way into the official Maritime Military Strategy.)

Faulty attribution notwithstanding, the Chinese is clearly drawn to Mahanian notions of sea power when forecasting how India will approach its maritime environs. Zhao Bole, a professor of South Asian studies at Sichuan University, places these claims in a more concrete geopolitical context. Argues Zhao, four key geostrategic factors have underwritten India's rise. First, India and its surrounding areas boast a wealth of natural resources. Second, India is by far the most powerful country in the Indian Ocean region. Third, the physical distance separating the United States from India affords New Delhi ample geopolitical space for maneuver.

Fourth, India borders economically dynamic regions such as the Association of Southeast Asian Nations (ASEAN) states and China. Zhao quotes Nehru and K. M. Panikkar to prove that Indian politicians and strategists have long recognized these geopolitical advantages and that they have consistently evinced the belief that India's destiny is inextricably tied to the Indian Ocean. However, due to India's insistence on taking a third way during the Cold War superpower competition, New Delhi was content to focus on its own sub continental affairs. In the 1990s, though, Zhao argues, India sought to shake off its nonaligned posture by increasing its geopolitical activism in Southeast Asia under the guise of its "Look East" policy.

According to Zhao Gancheng, New Delhi leveraged its unique geographic position to make Southeast Asia - an intensely maritime theater - a "breakthrough point" (...), particularly in the economic realm. In the twenty-first century, Zhao argues, the Look East policy has assumed significant strategic dimensions, suggesting that India has entered a new phase intimately tied to its great-power ambitions. While acknowledging that the underlying strategic logic pushing India beyond the subcontinent is compelling, Zhao worries that Indian prominence among the ASEAN states could tempt the United States to view India as a potential counterweight to China.

To Chinese observers, these broader geopolitical forces seem to conform to the more outward-looking Indian maritime strategy on exhibit in recent years, and they tend to confirm Chinese suspicions of an expansive and ambitious pattern to India's naval outlook. Zhang Xiaolin and Qu Yutao divide the evolution of Indian maritime strategy, particularly with regard to its geographic scope, into three distinct phases:

1. Offshore defence (...) (from independence to the late 1960s)
2. Area control (...) (from the early 1970s to the early 1990s)
3. Open-ocean extension (...) (from the mid-1990s to the present).

During the first stage, the navy was confined to the east and west coasts of India and parts of the Arabian Sea and Bay of Bengal in support of ground and air operations ashore. The second phase called for a far more assertive control of the Indian Ocean. Indian strategists, in this view, divided the Indian Ocean into three concentric rings of operational control. First, India needed to impose "complete or absolute control" over three hundred nautical miles of water out from India's coastline to defend the homeland, the exclusive economic zone, and offshore islands.

Second, the navy had to exert "moderate control" over an ocean belt extending some three to six hundred nautical miles from Indian coasts in order to secure its sea lines of communications and provide situational awareness. Finally, the navy needed to exercise "soft control," power projection and deterrent capabilities, beyond seven hundred nautical miles from Indian shores.

Chinese analysts differ over the extent of Indian naval ambitions in the twenty-first century. But they concur that India will not restrict its seafaring endeavors to the Indian

Ocean indefinitely. Most discern a clear transition from a combination of offshore defence and area control to a blue -water offensive posture.

One commentator postulates that India will develop the capacity to prevent and implement its own naval blockades against the choke points at Suez, Hormuz, and Malacca. Unsurprisingly, the prospect that India might seek to blockade Malacca against China has attracted substantial attention. One Chinese analyst, using language that would have been instantly recognizable to Mahan, describes the 244 islands that constitute the Andaman-Nicobar archipelago as a "metal chain" (...) that could lock tight the western exit of the Malacca Strait.

Zhang Ming further argues that "once India commands the Indian Ocean, it will not be satisfied with its position and will continuously seek to extend its influence, and its eastward strategy will have a particular impact on China." The author concludes that "India is perhaps China's most realistic strategic adversary".

While they pay considerable attention to the potential Indian threat to the Malacca Strait, Chinese observers also believe the Indian sea services are intent on:

1. Achieving sea control from the northern Arabian Sea to the South China Sea
2. Developing the ability to conduct SLOC defence and combat operations in the areas above
3. Maintaining absolute superiority over all littoral states in the Indian Ocean
4. Building the capacity for strategic deterrence against outside naval powers
5. Amassing long-range power-projection capabilities sufficient to reach and control an enemy's coastal waters in times of conflict
6. Fielding a credible, sea-based, second-strike retaliatory nuclear capability
7. Developing the overall capacity to "enter east"

into the South China Sea and the Pacific, "exit west" through the Red Sea and Suez Canal into the Mediterranean, and "go south" toward the Cape of Good Hope and the Atlantic.

Clearly, the Chinese foresee the emergence of a far more forward-leaning Indian Navy that in time could make its presence felt in China's own littoral realm. Moreover, the Chinese uniformly believe that New Delhi has embarked on an ambitious modernization program to achieve these sweeping aims. Interestingly, some have pointed to America's apparent lack of alarm at India's already powerful navy. This quietude, they say, stands in sharp contrast to incessant U.S. concerns over the People's Liberation Army Navy (PLAN), representing a blatant double standard.

In any event, China's assessments of Indian capabilities and its emerging body of work tracking India's technological and doctrinal advances is indeed impressive. For instance, Modern Navy, the PLAN's monthly periodical, published a ten-month series on the Indian Navy beginning in November 2005. Subjects of the articles ranged widely, from platforms and weaponry to basing and port infrastructure. Not surprisingly, given the decades-long debate within China surrounding its own carrier acquisition plans, India's aircraft carriers have attracted by far the most attention.

A number of Chinese analysts, however, hold far less alarming, if not sanguine, views of India's rise. The former Chinese ambassador to India, Cheng Ruisheng, argues that policy makers in Beijing and New Delhi have increasingly abandoned their antiquated, zero-sum security outlooks. Indeed, Cheng exudes confidence that improving U.S.-Indian ties and Sino-Indian relations are not mutually exclusive, and thus he holds out hope for a balanced and stable strategic triangle in the region.

Some Chinese speculate that India's burgeoning friendships with a variety of extraregional powers, including the United States and Japan, are designed to widen India's room for maneuver in an increasingly multi-polar world without forcing it to choose sides. As Yang Hui asserts, "India's actions smack of 'fence-sitting.' This is a new version of non-alignment." On balance, then, strategic continuity might prevail over the potentially destabilizing forces of change. Even those projecting major changes in the regional configuration of power seem confident that India's rise will neither upend stability nor lead automatically to strategic advantages for New Delhi.

To be sure, a small minority in China believes that an increased Indian presence in the Indian Ocean would generate great-power "contradictions" that could in time lead New Delhi to displace the United States as the regional hegemony, consistent with more forceful conceptions of an Indian Monroe Doctrine. But a far more common view maintains that growing Indian sea power will likely compel Washington and other powers in Asia to challenge or counterbalance New Delhi's position in the Indian Ocean region. Structural constraints will tend to act against Indian efforts to wield influence beyond the Indian Ocean.

Zhao Gancheng, for example, argues that China's firmly established position in Southeast Asia and India's relative unfamiliarity with the region will prevent New Delhi from reaping maximum gains from its Look East policy. On the strictly military and technological levels, some Chinese analysts believe that Indian naval aspirations have far outstripped the nation's concrete capacity to fulfill them. Noting that increases in the defence budget have consistently outpaced the annual growth rate of India's gross domestic product, Li Yonghua of Naval and Merchant Ships derides India's ambition for an oceangoing naval fleet as a "python swallowing an elephant".

Similarly, Zhang Ming identifies three major deficiencies that cast doubt on India's ability to develop a fleet for blue-water combat missions. First, India's current comprehensive national power simply cannot sustain a "global navy" and the panoply of capabilities that such a force demands.

Second, India's long-standing dependence on foreign technology and relatively backward industrial base will severely retard advances in indigenous programs - especially plans for domestically built next-generation aircraft carriers.

Finally, existing Indian Navy surface combatants are unequal in both quantitative and qualitative terms to the demands of long-range fleet operations. In particular, insufficiently robust air-defence constitutes the "most fatal problem" for future Indian carrier task forces.

Interestingly, key aspects of Zhang's critique apply equally to the PLAN today. This brief survey of Chinese perspectives suggests that definitive conclusions about the future of Indian sea power would be premature. On the one hand, evocative uses of Mahanian language and worst-case extrapolations of Indian maritime ambitions certainly represent a sizable geopolitically minded school of thought in China.

On the other, the Chinese acknowledge that India may not be able to surmount for years to come the geopolitical and technological constraints it confronts. Such mixed feelings further suggest that Sino-Indian maritime competition in the Indian Ocean or the South China Sea is not fated. Neither side has the credible capacity-yet-to reach into the other's nautical backyard. At the same time, the broader geostrategic climate at the moment favors cooperation.

There should be ample time - until either side acquires naval forces able to influence events beyond its own maritime domain, and as long as New Delhi's and Beijing's extraregional aims remain largely aspirational - to shape mutual threat perceptions through cooperative efforts.

AN UNCERTAIN GEOMETRY

This initial inquiry into the maritime geometry of the Indian Ocean region suggests that conditions are auspicious for shaping a mutually beneficial maritime relationship among India, China, and the United States. For now, New

Delhi seems at once sanguine about its maritime surroundings and conscious that it lacks the wherewithal to make good on a muscular Monroe Doctrine. While in principle India asserts regional primacy, much as James Monroe's America did, it remains content to work with the predominant naval power, the United States, in the cause of maritime security in South Asia. If nothing else, this is a matter of expediency.

It is worth noting, however, that there is little prospect that India will join the United States to contain Chinese ambitions in the Indian Ocean as Japan joined the United States to contain Soviet ambitions. India's independent streak, codified in its policy of nonalignment, predisposes New Delhi against such an arrangement. Nor does India resemble Cold War-era Japan, dependent on an outside power to defend it against an immediate, nearby threat to maritime security, and indeed national survival.

The geographic conformation of Japan's threat environment significantly heightened the urgency of a highly alert strategic posture. The Japanese archipelago closely envelops Vladivostok, home to the Soviet Union's Pacific Fleet and the base from which commerce-raiding cruisers had harassed Japanese trade and military logistics during the Russo-Japanese War.

Tokyo had to develop the capacity to monitor Soviet hunter-killer submarines lurking in the Sea of Japan and to repel a massive amphibious invasion against Hokkaido. India, by contrast, enjoys two great oceanic buffers - the eastern Indian Ocean and the South China Sea - vis-à-vis China.

As a simple illustration, several thousand nautical miles separate the fleet headquarters of China's South Sea Fleet, located in Zhanjiang, Guangdong Province, from Vishakhapatnam, the eastern naval command of the Indian Navy. Geography alone, then, constitutes a major disincentive for New Delhi to enlist prematurely in an anti-China coalition. For its part, Washington has not yet dedicated serious attention and energy to analyzing the future of Indian sea power or the likely configuration of great power relations in the Indian Ocean.

It remains hopeful that a durable strategic partnership with New Delhi will take shape. Should the three sea powers manage to draw in other powers with little interest in infringing on India's Monroe Doctrine or capacity to do so - say, Australia, an Indian Ocean nation in its own right, or Japan, which depends on Indian Ocean sea-lanes for energy security-the regional geometry could become quite complex. But the participation of such powers might also reduce the propensity for competition among the three vertices of the Sino-Indian-U.S. triangle. A wider arrangement, then, warrants study in American strategic circles.

Also, as we have seen, China views India's maritime rise with equanimity for now, doubting both New Delhi's capacity and its will to pose a threat to Chinese interests in the region. American hopes and Chinese complacency may not add up to an era of good feelings in South Asia, but they may form the basis for cooperative relations in the near to middle term. But this inquiry also suggests that the opportunity to fashion a tripartite seagoing entente may not endure for long. If India succeeds in building powerful naval forces, it may - like Cleveland's or Roosevelt's America - set out to make the Indian Ocean an Indian preserve in fact as well as in principle.

If so, China would be apt to take a more wary view of Indian naval ambitions, which would seem to menace Chinese economic, energy, and security interests in South Asia. Its hopes for a strategic partnership dashed, the United States might reevaluate its assumptions about the viability of a consortium of English-speaking democracies. This too would work against a cooperative strategic triangle. Maritime security cooperation, then, is by no means foreordained. A host of wild cards could impel New Delhi toward a more forceful security doctrine. Should, say, the

United States use the Indian Ocean or the Persian Gulf to stage strikes against Iranian nuclear sites, New Delhi might see the need to expand its regional primacy at America's expense.

A failure of the U.S.-Indian civilian nuclear cooperation accord would have an unpredictable, if indirect, impact on the bilateral relationship, fraying Indian patience and potentially loosening this "side" of the strategic triangle. Similarly, if China began deploying ballistic missile submarines to the Indian Ocean, India might redouble its maritime efforts, working assiduously on antisubmarine warfare and its own undersea nuclear deterrent. Competition, not cooperation, could come to characterize the strategic triangle - perhaps giving rise to some other, less benign regional geometry.

6

Anglo-American Interests In The Indian Ocean

When men such as Robert Clive and Warren Hastings returned to England, instead of settling down to enjoy a quiet life of ease and privilege, as they had dreamed, they became the object of "scandal." Their past behavior was scrutinized by parliament through investigations and hearings. Their reputations were battered and even destroyed. Parliament attempted to redress the situation of unruly and exploitative behavior abroad by passing successive laws to regulate and curtail the activities of the East India Company. So much so that by 1857, the Crown assumed direct control over Indian affairs.

Dirks therefore insists that the "scandals" of the late eighteenth century were used by British leaders as a means of gradually justifying and formalizing British power over India: He laments the fact that the British did not use these scandals to recognize "the scandal of empire itself." However, the author does not offer viable alternatives to the predicament that the British government faced. What else could they do? Not taking any action would leave the field open to unscrupulous capitalists and adventurers (British and European), who would continue to use the preponderant wealth and technology of the West to brutalize the local people even more—which would have been far more "scandalous" indeed. Dirks devotes much attention to one of the most sensational spectacles in the history of the British parliament: the impeachment trial of the governor of Bengal, Warren Hastings.

The trial lasted an astonishing nine years—from 1788 to 1795—and was conducted primarily at the behest of the Whig parliamentarian, Edmund Burke. Dirks insists that, although Burke meant well, the trial unintentionally encouraged further British excursions abroad; it led to the gradual establishment of the false idea that the British government, rather than employees of the East India Company, could do a better job of controlling private excesses. The impeachment proceedings against Hastings ultimately failed, but the British Empire was soon considered to be a sacred enterprise. Thus, the trial of Warren Hastings served in the long run to legitimize the British Empire.

Dirks explains how in subsequent generations, the scandals of the early part of the empire were erased from public memory. Instead, the very men who were ignominious in their day were presented as the heroic founders of the British Empire. The underlying mission of The Scandal of Empire is to explain that the excesses which occurred in the early phases of imperial growth cannot simply be ignored as anomalies. Instead, they define empire itself. Dirks laments the fact that imperialism is not yet considered wholly bad—as are slavery and fascism. He is trying to transform the consensus among imperial historians of viewing empires as consisting of a mixture of both positive and negative legacies. (On the one hand, European imperialists spread Christianity, education, health care, free markets, Western ideals of self-rule, the rule of law, and the concept of individual rights; on the other hand, imperialists were also responsible for war, profiteering, and oppression.) Instead, Dirks insists that there are no justifications for imperialism.

Dirks further maintains that much of the wealth that was accumulated via the British Empire resulted in the

Industrial Revolution; thus, the prosperity of British society as a whole (not just that of a few excessive governors) is in itself a scandal. The author therefore echoes the famous thesis popularized in 1944 by the socialist historian Eric Williams in Capitalism and Slavery: the Industrial Revolution was made possible by the accumulated savings of imperialists and the wealth of the West was due to the rapacious conquest of foreigners. Therefore, according to Dirks, it is not only the British Empire that is illegitimate but also the global capitalist economy to which the empire contributed. Despite Dirks's diligence in collecting as many of the scandalous occurrences of the late eighteenth century that he can find, he fails to advance our understanding.

The primary problem with the work is that "scandal" is insufficient as a means of characterizing centuries of rule; it is simply too monolithic and sweeping. Furthermore, one can well ask: what public or social institution has not been beset by scandal? We could do a similar study of any social institution and highlight all the times it failed to measure up to its founding ideals: even institutions that are ostensibly benign, such as the family, the university, humanitarian relief agencies, and others could be seen to be riddled with "scandal." That the imperial age had scandals is rather obvious.

Nor is evidence of these scandals sufficient to disqualify all the humanitarian efforts made in the name of empire. For every instance of appalling greed and abuse, one can find an example of stirring altruism; there are abundant examples in the history of the British Empire of men and women willing to lay down their lives for no other reason than to do what they believed would greatly benefit their fellow man.

Ultimately, Dirks fails because he overreaches. Furthermore, Dirks makes a profoundly anachronistic argument. It was the very "scandals" he laments which became watershed moments in Western thought and debate: these discussions raised social standards. Burke's impassioned pleas on behalf of the Indian people established an unprecedented ideal among imperialists.

Which other empire before Burke's time debated how well its subjects were being treated? Moreover, which other empire established the precedent that imperial rule was to be conducted in order to benefit those governed? Hence, we could very well interpret all those same scandals as a testament to how increasingly scrupulous the British public became. In the late eighteenth century, in tandem with these "scandals," there occurred an evangelical revival which fueled the movement to abolish slavery. By the early nineteenth century, the British paved the way for the greatest humanitarian reform movements the world had ever seen: the suppression of slavery, the expansion of the franchise, the rights of women, the reform of education and the military, and the birth of nursing as a profession.

Hence, the fact tha. the behaviour of some imperialists aroused scandal can just s well be interpreted as the result of a highly sophisticated, highly developed social conscience. Thus, these debates may be considered the fountainhead of the Western regard for individual rights and liberties. Yet, even more disturbing than this faulty and anachronistic analysis is Dirks's frequent references to contemporary issues. By assaulting the British Empire, he is really trying to assault the current policy of the American government in Iraq, which he believes to be imperialistic. The author's infusion of passionate contemporary polemics in a historical monograph is highly distracting and unbecoming a senior scholar.

Unfortunately, for all these reasons, The Scandal of Empire will not likely have much impact within imperial historiography. Historians have long debunked the idea that any empire—whether Western or Eastern, Christian or Muslim—can be viewed as either exclusively bad or exclusively good. Empires, like nations, are far too complicated, encompass too many time frames, and contain too many diverse occurrences to be reduced to simple characterizations—especially since former empires form the basis of much that our contemporary global society has decided to preserve.

While Brian W. Richardson's Longitude and Empire: How Captain Cook's Voyages Changed the World does hot surfer from the more egregious simplifications of Dirks's work, it too contains an unsustainable thesis. Richardson, a librarian at Windward Community College in Hawaii, examines three voyages undertaken for scientific purposes by Cook into the South Pacific from 1768 to 1780. Richardson insists that although these travels did hot result in new geographic discoveries or conquests, Captain Cook can nonetheless be seen as a founder of the British Empire.

By applying scientific analysis and enlightened thought to his travels, Cook presented the world as navigable and knowable rather than mysterious, fearsome, wild, and vast. Thus, according to the author, Cook changed the European worldview and contributed to setting the intellectual foundations for imperial adventure. The author makes his claims based exclusively on a textual analysis of Cook's travel diaries, which were published and widely read in his day. Richardson is hot interested in discovering what Cook actually did or who he actually was: he uses the travel journals to extrapolate Cook's legacy. Richardson is on very solid ground when he demonstrates how Cook's travel diaries served as a model for those who ventured abroad.

Indeed, Cook was esteemed for clarifying locations of islands, updating information, and creating a mathematical and scientific description of the world's places. He raised the standard for meticulous depiction. Captain Cook's work was especially influential in presenting his discoveries with reference to specific coordinates. Cook's writings helped to span a genre of literature, the travelogue, which became immensely popular in the imperial age. Travel diaries fed the public's insatiable appetite for knowledge about the people and places being discovered. These writings also inspired many young men to undertake similar excursions.

Hence, Cook was indeed an exemplary traveler and reporter.

Richardson maintains that Cook's travels were also influential in the debates on the nature of the nation. The descriptions in his travels clearly linked peoples to a specific territory: hence, rather than the notion that states created territory, Cook's depictions revealed that territory created states. Furthermore, Cook believed that the civilizations he encountered were the result of their environment or their climate; their structures were not fixed or intrinsic but could be altered. He presented himself as an enlightened benefactor. As such, he became the prototype of the European who could improve the lot of his fellow man. His writings projected the enlightenment ideal that nations could be altered to conform to "universal" principles of decency, order, and efficiency.

However, Richardson overstates Captain Cook's influence as an imperial progenitor and his responsibility for a dramatic alteration in the European view of the world. While Cook wrote model travel diaries, he was not, in any substantial way, an innovator or one who "changed the world." The age of discovery and of meticulous explanation of the world and its inhabitants had commenced long before the voyages of Captain Cook—as far back as the late fourteenth century. For several hundred years, descriptions of native peoples and their customs, either native American or Africans, trickled into Europe. These descriptions too, under the overall rubric of "savage," described indigenous groups in meticulous detail.

Nor was the idea of travelling abroad in order to perform benevolent acts originally or exclusively an enlightenment ideal—and it was not given its greatest impetus by Captain Cook. Since ancient times, missionaries had migrated to far-flung areas of the world for what they conceived as humanitarian and noble ends. The missionary zeal which animated much travel throughout the empire in the nineteenth century was born less out of the conviction that science and reason could remake the world in its image than that souls needed to be saved. In many instances, this was an anti-enlightenment project, as missionaries saw themselves at odds with the current of western cultural imperialism and ruthless capitalist enterprise.

And finally, the empire was not bore out of any need the British had of "collecting" parts of the world or applying scientific and enlightenment ideals to their possessions: the British Empire was created by fits and starts and for a variety of reasons—defensive, economic, religious, and humanitarian. The British Imperial ideal did not precede the birth of the empire, but rather emerged after the empire had already been created. Due to the fact that the British elite Were steeped in classical literature, it was to ancient ideals of empire that they turned, in an attempt to understand and justify their possessions.

By the nineteenth century, it was not enlightenment or scientific notions of collecting the world that inspired the British public, but rather the idea that they were participating in a natural, longstanding, and inevitable tradition of great empires dating from the ancient world. The only novel feature from their perspective was that this time, rather than empire being Roman, Greek, Aztec, Spanish, Arab, Indian, Ottoman, African, or Chinese, the British awoke to the realization that, paradoxically, a small island nation with representative institutions was the preeminent world power. Thus, while Richardson does a good job of explaining the nature and scope of Cook's voyages, he does not contextualize these properly.

In essence, Cook was a popular figure who set high standards as a travel writer and inspired a genre of literature with much influence in the imperial age. However, he was hardly as seminal as the author would have us believe. The Indian Ocean region had a unique oceanic culture—one that spawned "a hundred horizons" and contained universal ideals. Bose insists that the fresh research conducted in both global and local history renders the nation an insufficient category in defining the actions and identities of colonized peoples. He rejects the "macro" view of empire: the idea that western rule was so potent that it entirely subsumed the cultures that it ruled. He is also dissatisfied with the "micro" view which focuses narrowly on the activities of local communities.

Instead, he presents the existence of an "interregional arena" as a model for understanding resistance to, and accommodation of, imperial power. In the Indian Ocean region, Bose maintains, there was, alongside nationalist ideals, a more universal and international anti-colonial fervor. The latter allowed various degrees of local control and autonomy, which better accommodated religious and ethnic differences. As a result, once the British departed, the Indian people pursued a nationalist model for their new state, which was not adept at fusing diverse cultures or beliefs.

Yet, despite the imposition of more rigid control than the Indian people had hitherto known, there was a global community within the Indian Ocean region throughout much of the heyday of empire. This consisted primarily of links to a worldwide trade network. Indian migrant and indentured laborers went to North America, Asia, and Africa. Trade flourished with African, Asian, and Middle Eastern nations—especially in commodities such as cloves, pearls, rice, rubber, and oil. Indian and Chinese capitalists travelled and exchanged goods throughout the Indian Ocean. Thus, the Indian Ocean region had a fluid, vibrant, and unique culture.

Moreover, the British inadvertently fostered this global network by establishing a European-style standing army in India. Indian troops travelled the globe on British missions; they were especially active in the belt from North Africa to East Asia. The number of Indian troops employed by the British swelled dramatically in World War I, The author skillfully illustrates that the soldiers in British employ adopted a sense of universal mission: they perceived themselves as assisting in the fight for liberty against tyranny; they were citizens of the world.

They developed neither a British imperial identity nor an Indian nationalist one; instead, they had solidarity with one another in the fight for large, universal ideals of human benevolence. Bose also uncovers a "diasporic patriotism" in the thought of political activists, pilgrims, scholars, and poets. Gandhi, for example, returned from his travels to Britain and South Africa determined to foster Hindu-Muslim harmony in his native land: "Gandhi saw no contradiction between territorial nationalism and extraterritorial universalism—which was often tinged with religious inspiration."

Muslim pilgrims also identified with a worldwide community of Islam, fostering a universal identity that contained within it the seeds of anti-colonial resistance. Such universalism, both Muslim and Hindu, was also disseminated by scholar-pilgrims like Khwaja Hasan Nizami and the poet, Rabindranath Tagore, who won the Nobel Prize in Literature in 1913. As a result of these activities, the culture of the Indian Ocean region undermined the Christian universalism propagated by many British imperialists by presenting a rival creed: Muslim universalism. The imperial age can therefore best be viewed as consisting of "competing universalisms."

Bose insists, there were many sources to the process of globalization and many different contributors: the peoples of the Indian Ocean played a seminal role, as partners with the Europeans, in establishing a global society. In the final analysis, Bose's A Hundred Horizons adds an important dimension to the imperial literature.

The scholar makes a persuasive case for looking at the imperial age through new lenses. He demonstrates that there were indeed regional identities that transcended borders. Furthermore, he reveals how powerful was the role of religious universal ideals in resisting British political and cultural hegemony.

This adds to the growing body of evidence on the way in which the British Empire contributed, both deliberately and inadvertently, to the establishment of a global era. According to common wisdom, the mantle of the British imperial legacy has passed to the United States of America. With the disintegration of the British Empire after World War II and the rise of America as one of the world's two superpowers during the Cold War and as the sole superpower since the collapse of the Berlin Wall, there is much discussion now as to whether or not America is an empire.

To what extent does America act in the same manner as the former British Empire did in its prime? Is the new world order substantially different from the time when the British had an empire upon which the sun never set? To this debate, Charles S. Maier adds remarkable erudition and a convincing interpretation. In Among Empires: American Ascendancy and Its Predecessors, the Leverett Saltonstall Professor of History at Harvard University contrasts the current American role in international affairs with the world's great empires across the ages. He concludes that America is not an empire in the traditional understanding of the word: the United States does not exert formal control over other nations, nor does America embark on a systemic campaign of conquest.

Yet, neither is the U.S. merely the insular and self-contained nation envisioned by the Founding Fathers. America currently rules much of the world by fostering cooperation in areas of common interest and by using moral suasion. Hence, the author believes that America is currently a "hegemon" rather than an empire. The nation has many traits exhibited by former empires but has not yet fully embarked upon a full-blown imperial course. This may be in part because we live in a different era: Maier suggests that the global influence America has through its military supremacy, economic might, technology, and spread of culture points to power that is "post-territorial."

America's hegemony is not necessarily confined within specific borders: the frontiers of empire are not as they were in previous centuries. Thus, in reaction to this kind of power, even American detractors, such as al Qaeda, aspire to create a post-territorial empire. The author explores the question of whether or not American hegemony has resulted in more or less violence in the world. He does not provide a definitive answer to this but emphasizes that "the lifeblood of empire is blood." Violence is an inevitable aspect of all empires: there is violence on the frontiers, violence from dissenters who resist imperial rule, and violence when the empire disintegrates into constituent parts. This does not mean that other forms of government are necessarily less violent.

Maier then proceeds to trace the three primary phases of American ascendancy. First, America displaced Britain in its industrial capacity and military strength. This occurred as a result of the mass-factory system established by Henry Ford at Highland Park and later by the development of nuclear weapons. Thus, by the end of World War II, the British Empire had been eclipsed. The second phase of American ascendancy occurred during the Cold War. The American government provided financial assistance, through the Marshall Plan to Europe, in exchange for forging an alliance against the Soviet Union. In this manner, America exercised great influence over West European nations.

The U.S. also created a powerful "empire of production" whose goods and culture were admired and emulated by Western nations. In this period, Americans were embroiled in a series of distant wars in order to curtail the influence of the Soviet Union; thus their frontier was often perceived to be very far from American shores. And finally, in the third stage of ascendancy, the American economy was transformed into one of consumption—one which specializes in service and distribution occupations and allows for other nations to establish industries that serve the American domestic appetite for goods.

According to Maier, it is too soon to tell whether this is a path that will embroil America into deeper, formal control over other nations or whether this is a temporary occurrence in response to what is perceived as a national threat. Maier expresses much admiration for the U.S., while at the same time fearing that American domestic institutions will be gradually subverted if the nation becomes a formal empire. Charles Maier thus provides a crisp and lucid synthesis of America's rise to hegemonic influence. He presents the growth of American power in the context of a world historically and consistently inhabited by empires.

Even if we examine the current area, which is under intense international scrutiny and often presented as the victim of Western intervention—the Middle East—we cannot escape confronting the great shadow of former empires; Maier draws attention to the fact that the Iraqi people, in essence, have existed for fifty centuries largely within a series of empires. Hence, the author presents a global and comparative approach to the rise of American hegemony that is grounded in a profound and commanding knowledge of the past. This historical and global context is vital in an assessment of empire—whether American or British. Empires have been, from ancient times, a consistent feature of every civilization.

In the twentieth century, even the creation of alternative international structures such as the League of Nations or the United Nations have not been able to withstand or supersede large constellations of power, which crystallize around, a preponderant force. Nor has the nation-state cured all the ills usually attributed to imperial domination: the nation-state has not been exempt from extremes of violence, intolerance, racial hatred, social discord, inequality, and oppression. Hence, it is futile to lament the fact that despite their ideals and their rhetoric, the large English-speaking nations have thus far failed to create merely small, self-governing republics and have not abstained from dominating other nations.

Both the British and Americans have been compelled gradually to expand their power in defence of their vital interests. Democracies, declared nineteenth-century English Radicals such as Richard Cobden and John Bright, do not fight wars and do not have empires. Yet, both the British and Americans who led the way in exemplary representative institutions have consistently been embroiled in both war and empire. Is this an inevitable trajectory—as with the Greeks and Romans? If so, then the supreme question for British and American imperial scholars is: what is the ultimate contribution to the progress of global civilization of Britain and America? And, are there any empires, in the history of the world, that have a better legacy?

Pakistan's weakness compounded U.S. difficulties in shoring up security in Afghanistan. Pakistan's ungoverned border region with Afghanistan harboured al Qaeda and Taliban militants working to overthrow the U.S.-backed administration in Kabul. Pakistani terrorists also threatened India: one such group was implicated in the dramatic November 2008 terrorist attacks in Mumbai. Without stronger Pakistani government efforts to suppress such groups and stop blatant attacks on India, New Delhi's retaliation with military and other actions would raise the spectre of a major confrontation between the two nuclear armed rivals.

Meanwhile, developments in the Middle East stalled prospects for advancing peace amid deep regional and global concerns over Iran's apparently active pursuit of nuclear weapons. Against this background, U.S. relations with the rest of the Asia-Pacific region seemed likely to be of generally secondary importance for U.S. policy-makers. The global economic crisis put a premium on close U.S. collaboration with major international economies, notably Asian economies like China and Japan, in promoting domestic stimulus plans, supporting international interventions to rescue failing economies and avoiding egregiously self-serving economic and trade practices that could prompt protectionist measures seen to encumber any early revival of world economic growth.

Apart from the deeply troubled Middle East-Southwest Asian region, the other major area of U.S. security concern in Asia is North Korea. Pyongyang climbed to the top of the Obama government's policy agenda through a string of provocative actions in 2009 culminating in North Korea's withdrawal from the Six Party Talks and its second nuclear weapons test in May. North Korea's first nuclear weapons test of 2006 represented a failure of the Bush administration's hard line approach in dealing with North Korea's nuclear weapons programme. In response, that U.S. administration reversed policy, adopting a much more flexible approach, including frequent bilateral talks with North Korean negotiators, within the broad framework of the Six-Party Talks seeking the denuclearization of the Korean peninsula.

Important agreements were reached but North Korea did not fulfill obligations to disable and dismantle plutonium-based nuclear facilities. The Obama government had seemed poised to use the Six Party Talks and bilateral discussion with North Korea in seeking progress in getting Pyongyang to fulfill its obligations. The escalating North Korean provocations in 2009 and the Pyongyang regime's strident defiance of UN Security Council resolutions and international condemnation compelled a U.S. policy review. Obama government leaders from the President on down also consulted closely with concerned powers, notably key allies Japan and South Korea, and China, in assuring a firm response from the UN Security Council in June that imposed sanctions in addition to those imposed after North Korea's first nuclear test and called for inspections of suspected weapons shipments to and from North Korea.

The United States also planned its own unilateral sanctions in order to pressure Pyongyang to halt the provocations and return to negotiations. Available evidence in mid-2009 showed considerable skepticism that negative and positive incentives from the United States and other concerned powers would lead to improvement in North Korea's behaviour. Few were optimistic that the crisis atmosphere would subside soon. Meanwhile, beginning in 2008, longstanding U.S. concerns with the security situation in the Taiwan Straits declined as the newly installed government of President Ma Ying-jeou reversed the pro-independence agenda of his predecessor in favour of reassuring China and building closer cross-strait exchanges.

The Obama administration indicated little change from Bush administration efforts to support the more forthcoming Taiwan approach and avoid U.S. actions that would be unwelcome in Taipei and Beijing as they sought to ease tensions and facilitate communication. The Obama administration and the strong Democratic majorities in both Houses of the Congress also gave high priority to promoting international efforts on the environment and climate change. Such efforts appeared ineffective without the participation of Asia's rising economies, notably China, the world's largest emitter of greenhouse gases.

An American approach of prolonged consultation and dialogue with China to arrive at mutually acceptable approaches to these issues seems likely. This chapter assesses salient strengths and weaknesses of the United States in Asia at the start of the Obama administration, and reviews the new U.S. government's approach to key U.S. allies in the region and other Asian powers. It then examines the U.S. administration's policies and approach to Southeast Asian and Asian regional organizations and groupings where the Association of Southeast Asian Nations (ASEAN) plays a leading role.

The findings of the assessment show that the United States remains in a strong leadership position in Asia. The Obama administration has a major crisis on its hands in North Korea; there is no certainty in mid-2009 of whether or how the crisis will be resolved. Elsewhere, the new U.S.

government seems intent on correcting some generally secondary shortcomings in the Bush administration's efforts in the region. Apart from a possibly significantly higher profile for Southeast Asia and Asian multilateralism in U.S. policy, the new U.S. government's policy actions seem to reflect adjustments in order to increase benefits for the United States rather than larger scale policy revisions and change.

STRENGTHS AND WEAKNESSES OF U.S. LEADERSHIP IN ASIA

Media and specialist commentary as well as popular and elite sentiment in Asia tended to emphasize the shortcomings of U.S. policy and leadership in Asia for much of the Bush administration years. Heading the list were widespread complaints with the Bush administration's hard line policy towards North Korea, its military invasion and occupation of Iraq, and assertive and seemingly unilateral U.S. approaches on wide ranging issues including terrorism, climate change, the United Nations and Asian regional organizations. The United States appeared alienated and isolated, and increasingly bogged down with the consequences of its invasion of Iraq and perceived excessively strong emphasis on the so-called "war on terrorism". By contrast, Asia's rising powers, and particularly China, seemed to be advancing rapidly.

China used effective diplomacy and rapidly increasing trade and investment relationships backed by China's double digit economic growth in order to broaden its influence throughout the region. China also carried out steady and significant increases in military preparations. This basic equation of Chinese strengths and U.S. weaknesses became standard fare in mainstream Asian and Western media. It was the focus of findings of many books and reports of government departments, international study groups and think-tanks authored often by well respected officials and specialists. The common prediction was that

Asia was adjusting to an emerging China-centred order and U.S. influence was in decline.

Over time, developments showed the reality in the region to be more complex. Japan clearly was not in China's orbit; India's interest in accommodation with China was very mixed and overshadowed by a remarkable upswing in strategic cooperation with the United States; Russian and Chinese interest in close alignment waxed and waned and appeared to remain secondary to their respective relationships with the West; and South Korea, arguably the area of greatest advance in Chinese influence at a time of major tensions in the U.S.-Republic of Korea relationship earlier in the decade changed markedly beginning in 2004 and evolved to a situation of often wary and suspicious South Korean relations with China seen today.

Former U.S. officials pushed back against prevailing assessments of U.S. decline with a variety of tracts underlining the U.S. administration's carefully considered judgement that China's rise was not actually having a substantial negative effect on U.S. leadership in Asia, which remained healthy and strong. They joined a growing contingent of scholars and specialists who looked beyond accounts that inventoried China's strengths and U.S. weaknesses and carefully considered other factors including Chinese limitations and U.S. strengths before making their overall judgments. Several commentators and think-tanks that had been prominent in warning of U.S. decline and China's rise revised their calculus to focus more on Chinese weaknesses and U.S. strengths.

What has emerged is a broad based and mature effort on the part of a wide range of specialists and commentators to more carefully assess China's strengths and weaknesses along with those of the United States and other powers in the region. The basic determinants of U.S. strength and influence in Asia seen in the recent more balanced assessments of China's rise and U.S. influence in Asia involve the following factors:

Security

In most of Asia, governments are strong, viable and make the decisions that determine direction in foreign affairs. Popular, elite, media and other opinion may influence government officials in policy towards the United States and other countries, hut in the end the officials make decisions on the basis of their own calculus. In general, the officials see their governments' legitimacy and success resting on nation-building and economic development, which require a stable and secure international environment.

Unfortunately, Asia is not particularly stable and most governments privately are wary of and tend not to trust each other. As a result, they look to the United States to provide the security they need to pursue goals of development and nation-building in an appropriate environment. They recognize that the U.S. security role is very expensive and involves great risk, including large scale casualties if necessary, for the sake of preserving Asian security. They also recognize that neither rising China nor any other Asian power or coalition of powers is able or willing to undertake even a fraction of these risks, costs and responsibilities.

Economics

The nation-building priority of most Asian governments depends importantly on export-oriented growth. Chinese officials recognize this, and officials in other Asian countries recognize the rising importance of China in their trade; but they all also recognize that half of China's trade is conducted by foreign invested enterprises in China, and half of the trade is processing trade—both features that make Chinese and Asian trade heavily dependent on exports to developed countries, notably the United States. In recent years, the United States has run a massive and growing trade deficit with China, and a total trade deficit with Asia valued at over US$350 billion at a time of an overall U.S. trade deficit of over US$700 billion.

Asian government officials recognize that China, which runs a large overall trade surplus, and other trading partners of Asia are unwilling and unable to bear even a fraction of the cost of such large trade deficits, that nonetheless are very important for Asian governments. Obviously, the 2008-09 global economic crisis is having an enormous impact on trade and investment. Some Asian officials are talking about relying more on domestic consumption but tangible progress seems slow as they appear to be focusing on an eventual revival of world trade that would restore previous levels of export oriented growth involving continued heavy reliance on the U.S. market.

Government Engagement and Asian Contingency Planning

The Obama administration inherited a U.S. position in Asia buttressed by generally effective Bush administration interaction with Asia's powers. It is very rare for the United States to enjoy good relations with Japan and China at the same time, but the Bush administration carefully managed relations with both powers effectively. It is unprecedented for the United States to be the leading foreign power in South Asia and to sustain good relations with both India and Pakistan, but that has been the case since relatively early in the Bush administration.

It is also unprecedented for the United States to have good relations with Beijing and Taipei at the same time, but that situation emerged during the Bush years and strengthened with the election of President Ma Ying-jeou in March 2008. The U.S. Pacific Command and other U.S. military commands and organizations have been at the forefront of wide ranging and growing U.S. efforts to build and strengthen webs of military relationships throughout the region. In an overall Asian environment where the United States remains on good terms with major powers and most other governments, building military ties through education programmes, on-site training, exercises and other means enhances U.S. influence in generally quiet but effective ways.

Part of the reason for the success of these efforts has to do with active contingency planning by many Asian governments. As power relations change in the region, notably on account of China's rise, Asian governments generally seek to work positively and pragmatically with rising China on the one hand; but on the other hand they seek the reassurance of close security, intelligence and other ties with the United States in case a rising China shifts from its current generally benign approach to one of greater assertiveness or dominance.

NON-GOVERNMENT ENGAGEMENT AND IMMIGRATION

For much of its history, the United States exerted influence in Asia much more through business, religious, educational and other interchange than through channels dependent on government leadership and support. Active American non-government interaction with Asia continues today, putting the United States in a unique position where the American non-government sector has such a strong and usually positive impact on the influence the United States exerts in the region. Meanwhile, over forty years of generally colour-blind U.S. immigration policy since the ending of discriminatory U.S. restrictions on Asian immigration in 1965 has resulted in the influx of millions of Asian migrants who call America home and who interact with their countries of origin in ways that undergird and reflect well on the U.S. position in Asia.

No other country, with the possible exception of Canada, has such an active and powerfully positive channel of influence in Asia. In sum, the findings of these assessments of U.S. strengths show that the United States is deeply integrated in Asia at the government and non-government level. U.S. security commitments and trade practices meet fundamental security and economic needs of Asian government leaders and those leaders are very aware of this.

The leaders also know that no other power or coalition of powers is able or willing to meet even a small fraction of those needs. And Asian contingency planning seems to work to the advantage of the United States, while rising China has no easy way to overcome pervasive Asian wariness of Chinese longer term intentions. On balance, the assessments show that the Obama administration can work to fix various problems in U.S. policy in Asia with the confidence that U.S. leadership in the region remains broadly appreciated by Asian governments and unchallenged by regional powers or other forces.

RELATIONS WITH KEY ALLIES

Though relations with rising China and India and troubles with North Korea tend to get the lion's share of media and public attention in U.S. relations with Asia, U.S. policy-makers continue to give priority to key allies in the Asia-Pacific region—Japan, South Korea and Australia. In a regional environment undergoing great change in relative power relationships and economic development and turmoil, these partners share American perceptions of the world order based on many decades of close security cooperation and shared values. South Korea and especially Japan also provide the bases needed for the United States to overcome the tremendous and often unappreciated obstacle of great geographic distance in the Pacific Ocean so that U.S. forces are readily available to preserve stability in and U.S. access to Asia.

The Bush administration strongly emphasized U.S. relations with Japan and Australia. President Bush's relationships with Japanese Prime Minister Junichiro Koizumi and Australian Prime Minister John Howard were the closest between the United States and these countries in recent memory. The two leaders were undaunted by formidable domestic and foreign opposition as they aligned closely with the Bush administration's controversial policies in the war on terrorism and the invasion of Iraq.

The Bush administration had much less success in relations with the left-leaning South Korean government of Roh Moo Hyun. U.S.-South Korean relations reached a crisis point by 2004 over strong differences regarding how to deal with North Korea and various bilateral disputes related to U.S. bases and troops in South Korea. Persistent U.S. interchange dealing with some of the South Korean concerns and strong recognition by the Roh administration of South Korea's continuing need for a strong alliance with the United States saw the South Korean government take some significant steps to shore up the relationship with the United States. Despite widespread domestic opposition, South Korea deployed and for several years maintained the third largest troop commitment in Iraq.

The Roh government also pushed hard to reach a free trade agreement with the United States. U.S.-South Korean relations improved with the end of America's hard line policy towards North Korea in 2006, bringing U.S. policy more in line with that of South Korea. The election in 2007 of a conservative, Lee Myeung Bak, as South Korea's president opened the way to closer U.S. collaboration with a South Korean government even more strongly inclined to solidify relations with its American ally. The early appointments of the Obama administration of key officials dealing with Asian affairs, Secretary of State Hillary Clinton's maiden visit to East Asia in February 2009 and President Obama's initial meetings with Asian leaders suggested that relations with these key allies would figure prominently in the regional and global calculations of the new U.S. government.

Prominent appointments of those well versed in U.S. alliance relations with Asia include Jeffrey Bader as senior Asian affairs policy coordinator in the National Security

Council, James Steinberg as Deputy Secretary of State, former Marine General Wallace "Chip" Gregson as Assistant Secretary of Defence for Asian and Pacific Affairs, and Kurt Campbell as Assistant Secretary of State for East Asian Affairs. Secretary Clinton's visit to Asia placed Japan and South Korea first in her four stops. Her remarks repeatedly underlined the importance the new U.S. government places on relations with Japan and other Asian allies. President Obama reinforced this point during later meetings with Asian leaders.

Since Koizumi left office in 2006, Japan's ruling Liberal Democratic Party (LDP) has had a string of weak and unpopular prime ministers who fended off repeated calls by the political opposition for parliamentary elections. The elections will come in 2009; if the opposition wins, there could be some adjustment and probable reduction in the Japanese government's very close identification with the United States during the rule of the LDP. The domestic political turmoil in Japan compounds difficulties caused by the major economic decline suffered by the export oriented Japanese economy as a result of the global economic crisis beginning in 2008. Despite these weaknesses, Japan remains the second largest economy in the world with great technological achievement and modern and capable armed forces.

U.S. leaders count on Japan to work with the United States in dealing with the economic crisis, in managing and reducing the threats posed by North Korea's nuclear weapons and other provocations and in sustaining regional peace through contingency plans guarding against possible assertive actions by a rising China and other potential sources of regional instability. Japan has developed and has shown strong willingness to share its leading expertise and extensive experience in environmental and climate change matters—key priorities of the new U.S. government.

South Korea's conservative president also is unpopular though his mandate is secure until his tenure ends in 2012.

Despite its comparatively small size, South Korea is among the world's top 15 economies. Its export oriented industries have been heavily impacted by the global economic crisis; its overall economic weight warrants inclusion in the Group of 20 and other groupings used by U.S. policy-makers to deal with the economic crisis. South Korea is even more important for U.S. policy dealing with nuclear North Korea and cooperating with the United States on the maintenance of 26,000 U.S. military personnel in the country. The shift in Bush administration policy towards greater flexibility on North Korea upset some Japanese officials and commentators who had been generally supportive of the previous hard line approach.

The Bush government also disappointed Japan by agreeing to remove North Korea from its list of terrorist supporting countries despite Pyongyang's refusal to account for Japanese abducted by North Korean agents. Media reports showed persisting antagonism between chief U.S. negotiator Christopher Hill and his Japanese counterparts over how to deal with North Korea. Meanwhile, with the shift in U.S. policy towards greater American collaboration with North Korea, the conservative Lee Myeong Bak government found itself as the focal point of North Korean antagonism and aggression. It sought support from the United States to buttress its position vis-a-vis North Korean pressure.

The Obama administration inherited this somewhat troubling legacy against the background of more deeply rooted Japanese and South Korean concerns about the security and economic policies of a Democratic administration. Republican candidate John McCain made a stronger point than the Obama campaign of a firm stance in support of Japan and South Korea in the face of North Korean and other threats. McCain also stressed strong commitment to free trade and open U.S. markets for Asian exports. The Obama campaign rhetoric was more reserved about free trade, notably opposing the Korea-U.S. Free

Trade Agreement. Its emphasis on reaching out to negotiate with adversaries also raised some uncertainties about the future American posture towards North Korea. A salient concern among officials in Japan and South Korea was that North Korea would outmanoeuvre an unseasoned U.S. administration that is looking for some early diplomatic success or that the Obama administration would seek an arrangement under which the United States would accept a nuclear armed North Korea as long as it does not proliferate.

There also was the more immediate concern that the United States, given its various distractions and preoccupations, would allow the North Korean situation to drift. A key question was whether or not the United States is prepared to push hard to press North Korea to meet its commitments in the Six-Party Talks, to disable and dismantle its plutonium producing facilities and to press Pyongyang further in order to continue towards full and verifiable denuclearization.

The appointment of former U.S. ambassador to South Korea Stephen Bosworth as the State Department's special representative for North Korea seemed to show U.S. resolve to ease Japanese and South Korean concerns. North Korea's subsequent military provocations appeared to overwhelm past concerns as the United States, Japan and South Korea worked closely together and with China and other concerned powers in devising negative and positive incentives sufficient to calm the North Korean provocations and resume the process of negotiations directed towards achieving a denuclearized Korean peninsula.

Adding to the mix of strategic uncertainties are anxieties, particularly in Japan, over a perceived tendency in the latter part of the Bush administration for the United States to give priority to interaction with rising China rather than longstanding U.S. allies. Presidential candidate Hillary Clinton initially appeared to neglect Japan in favour of China. More recent discussion of a U.S.-China "G-2" collaboration to deal with the economic crisis and related issues voiced by such influential individuals as Henry Kissinger, Zbigniew Brzezinski, C. Fred Bergsten and Robert Zoellick underline a persisting Japanese concern that the United States may seek improved U.S.-China relations at Tokyo's expense.

South Korea is particularly anxious to see modifications of Obama campaign positions against the U.S.-Korean Free Trade Agreement. Though media reports sometimes hint at ways the accord could be reached without fundamental renegotiation, the steps required remain formidable and how they would proceed amid the many other preoccupations of U.S. economic and trade policy-makers remains unclear. Meanwhile, both South Korea and Japan have issues with the United States over salient alliance questions. In Seoul, influential voices in the conservative government want to reassess the U.S.-South Korean agreement negotiated by its predecessor that transfers operational control of wartime command from U.S. to Korean forces, scheduled for 2012.

The continued stalemate between the United States and Japan over the relocation of the Futenma Air Station in Okinawa is seen as part of a broader failure to implement the roadmap for realignment of U.S. forces in Japan, agreed by the two countries in 2006. The Australian government of Prime Minister Kevin Rudd showed its independence from the Bush administration in markedly shifting Australian policy in favour of the Kyoto Protocol and in withdrawing Australian forces from Iraq. However, alliance cooperation with the United States remained broad and deep, with Australian forces continuing to fight alongside Americans in Afghanistan.

American interests continued to be well served by Australia's leading role in sustaining stability among the crisis-prone countries of East Timor and in the South Pacific. The United States and Australia collaborated in trilateral security cooperation with Japan. And Australia's improvement of security relations with Indonesia and South Korea seemed to meet with U.S. approval. Close Australian-U.S. economic relations grew as a result of the U.S.-Australian Free Trade Agreement of 2005. China's emergence as Australia's top trading partner and consumer of Australian resources prompted speculation that Australia was less willing than in the past to support U.S. positions on Taiwan. Australia also was said to be reluctant to join the United States and Japan in support for stronger demonstrations in military exercises and high-level security consultations involving India as a strategic partner of the three allies.

RELATIONS WITH CHINA: POSITIVE BUT FRAGILE EQUILIBRIUM

U.S.-China relations during the first decade of the twenty-first century evolved towards a positive equilibrium that appears likely to continue into the near future. Both the U.S. and Chinese administrations have become preoccupied with other issues and appear reluctant to exacerbate tensions with one another. Growing economic interdependence and cooperation over key issues in Asian and world affairs reinforce each government's tendency to emphasize the positive and pursue constructive relations with one another. The positive stasis provides a basis for greater cooperation over economic and security interests and issues.

At the same time, differences in strategic, economic, political and other interests have remained strong throughout the period and represent substantial obstacles to further cooperation between the two countries. Policy-makers in both countries continue to harbour suspicions about each others' intentions. Specialists in China and the United States have identified a pattern of dualism in U.S.-China relations that has emerged as part of the ostensibly positive equilibrium in the post-Cold War period. The pattern involves constructive and cooperative engagement on the one hand and contingency planning or hedging on the other. It reflects the mix noted above of converging and competing interests and prevailing leadership suspicions and cooperation.

Chinese and U.S. contingency planning and hedging against one another sometimes involves actions like the respective Chinese and U.S. military build-ups that are separate from and develop in tandem with the respective engagement policies the two countries pursue with each other. At the same time, dualism shows as each government has used engagement to build positive and cooperative ties while at the same time seeking to use these ties to build interdependencies and webs of relationships that have the effect of constraining the other power from taking actions that oppose its interests. Differences between the two countries continue out of the limelight.

They are managed in over sixty dialogues and other high-level interaction between the two administrations. Secretary Clinton's visit to China in February 2009 and President Obama's meeting with Chinese President Hu Jintao at the G-20 summit in London in April followed the pattern in the latter years of the Bush administration in calling for deepening dialogue and development of "positive and constructive" relations.

Nonetheless, the differences between the two countries are readily apparent on the U.S. side, where they are repeatedly highlighted by U.S. media and interest groups concerned about various features of Chinese governance and practice, and where the majority of Americans give an unfavourable rating to the Chinese government.

They are less apparent in the more controlled media environment of China, though Chinese officials and government commentaries make clear their strong opposition to U.S. efforts to support Taiwan and to foster political change in China, as well as key aspects of U.S. alliances, security presence around China's periphery and positions on salient international issues ranging from the military use of space to fostering democratic change. The positive features of the relationship tend to outweigh the negatives for three reasons. First, both governments gain from cooperative engagement—the gains include beneficial economic ties, as well as cooperation over North Korea, terrorism, Pakistan and even Taiwan.

It also includes smaller progress on Iran and even less on Sudan and Myanmar. Second, Washington and Beijing recognize that, because of ever closer U.S.-China interdependence, focusing on negative aspects in U.S.-China relations would be counter productive to their interests. Third, U.S. and Chinese leaders recognize that, because of other major policy preoccupations they both have, focusing on negative aspects in U.S.-China relations would also be counter productive to their interests. In sum, it seems fair to conclude that the recent U.S. relationship with China rests upon a mutual commitment to avoid conflict, cooperate in areas of common interest and prevent disputes from shaking the overall relationship.

Against this background, the Obama government seems most likely to advance relations with China in small ways. It probably will show sufficient resolve to avoid conflict with China over trade, currency, environmental security, Taiwan, Tibet human rights and other issues that appear counterproductive for what seem to be more important U.S. interests in preserving a collaborative relationship with China and avoiding frictions with such an important economy at a time when international economic cooperation is of paramount importance.

Those in the United States who seek to give greater prominence to differences with China seem overwhelmed for now, particularly by the salience of the global economic crisis and the perceived U.S. need to be seen to cooperate with China in restoring international economic confidence.

Events in China or U.S.-China relations could bring their issues to the fore, as they did during Chinese crackdown on dissent and violence in Tibet in 2008. In the recent past, events such as China's efforts to purchase a U.S. oil company during a period of rising gasoline prices in the United States and product safety issues with Chinese consumer goods exported to the United States saw spikes of anti-China media commentary, congressional commentaries and investigations and other public discussion that damaged China's image with the American public. The U.S. administration remained on the sidelines in those instances as it pursued private dialogues with the Chinese government, preserving the positive but still fragile equilibrium in U.S.-China relations.

RELATIONS WITH INDIA

The mix of areas of convergence and areas of divergence is more positive in the case of U.S. relations with the other fast rising Asian power, India. Though the details of the Obama administration's policies towards India have not yet been clearly articulated, it seems likely that the positive momentum in U.S.-Indian relations begun in the Clinton administration and greatly advanced in the Bush administration will be continued. The strategic uncertainty that lies at the base of the respective "hedging" policies of the United States and China has little place in U.S.-Indian relations.

The wide ranging U.S.-Indian defence cooperation emerged and developed strongly at the turn of the decade and predates the negotiation and ratification of the U.S.-Indian civilian nuclear agreement—the centerpiece of the Bush administration's effort to promote India's development and world status. Since 2002, the United States and India have held a series of unprecedented and increasingly substantive combined exercises involving all military services. These have included the use of the Indian Air Force's advanced Russian SU-30MKI aircraft over U.S. territory, Special Forces training in mountains near the China-India border and annual naval exercises in the Indian and Pacific oceans.

The scope of arms sales has grown, with the United States welcoming Indian consideration of requests for advanced fighter aircraft and command-and-control, early warning and missile defence equipment. The U.S. government also supported Israel's sale of the U.S.-Israeli Phalcon airborne early warning system to India, while it worked hard to prevent the transfer of such advanced equipment from Israel to China. By the time of President Bush's 2006 visit to India, U.S. leaders were routinely highlighting the civilian nuclear accord as reflecting a core U.S. interest in working to help India become a major world power in the new century.

Washington conducted relations with New Delhi under the rubric of three major dialogue areas: strategic (including global issues and defence), economic (including trade, finance, commerce and the environment) and energy. As supporters of improved U.S.-Indian relations, the Indian-American caucus represented the largest of all country-specific caucuses in Congress. Multifaceted U.S.-India cooperation included close coordination in anti-terrorism efforts; in advancing the sales of U.S. high technology dual-use goods to India; and in major modifications of existing U.S. non-proliferation policy and law in order to allow for approval by the Congress and relevant international organizations of the U.S.-India civilian nuclear agreement.

Significant differences and limitations remained. India's economic reforms and greater conformity to trends in economic globalization reinforced growth and attracted U.S. investment and trade. However, problems included grossly inadequate infrastructure, widespread poverty and weak public services in health care, education, electric power and water supply. The Indian government also vacillates in its commitments to moving the economy in the direction of the free market. Excessive government control and bureaucracy are often cited in American complaints about doing business in India.

India aligns strongly with the poor developing countries against requirements sought by the United States and other developed countries that call upon India to support greater trade liberalization in negotiations at the World Trade Organization and other venues, and support emissions reduction in various climate change negotiations. Though India has made a strategic choice in pursuing a closer alignment with the United States, Japan and other developed countries, it also sustains a strong commitment to strategic independence in world affairs. Its improved relations with China and Pakistan are generally welcomed by the United States, though its close relations with Iran and Myanmar seem at odds with U.S. policy concerns.

RELATIONS WITH SOUTHEAST ASIA AND REGIONAL GROUPS FOCUSED ON ASEAN

Hillary Clinton's visit to Jakarta was clearly the public relations highlight of her first trip to Asia as Secretary of State. Official comment, media coverage and public assessments were positive about the Secretary and what her visit symbolized for increased U.S. attention to Indonesia and Southeast Asia more broadly. The Indonesians were also excited about the prospect of a "home coming" to Indonesia by President Barack Obama in connection with the annual Asia Pacific Economic Cooperation (APEC) meeting held in November 2009.

As a youth, Obama lived in Indonesia for several years. The Clinton visit helped to overcome some of the Southeast Asian ambivalence and wariness over Washington's episodic high-level attention and often offensive unilateral attitudes. The Secretary also built positive momentum for U.S. relations with the region by:

1. Highlighting Indonesia's status as a regional power and a vibrant and tolerant Muslim state.
2. Signaling U.S. interest in signing ASEAN's Treaty of Amity and Cooperation (TAC), which was eschewed by previous U.S. administrations even though the treaty has great symbolic importance in Southeast Asia and has been signed by China, Japan, Australia and other

powers?

3. Pledging to regularly attend the annual ASEAN Regional Forum (ARF) Ministerial Meeting in Southeast Asia — a contrast to previous Secretary of State Condoleezza Rice who missed two out of four of the annual meetings.

4 Indicating tactical flexibility in pressing for greater respect for human rights in Myanmar. The flexibility might open the way to closer U.S. cooperation with ASEAN which has been hampered by Myanmar's membership in the group and U.S. policies sanctioning and isolating the ruling junta.

Nevertheless, U.S. policy concerns regarding Southeast Asia remain less important than the range of issues seen in America's relations with China, Korea and Japan in Northeast Asia, and the array of interests and issues at stake in U.S. relations with India and neighbouring Pakistan and Afghanistan in South Asia. In part this lower priority prevails because U.S. policy-makers and opinion leaders generally see less at stake for American interests in Southeast Asia than in the other areas.

The U.S. military presence, trade and economic relations and Great Power politics inevitably give pride of place to Northeast Asia in American calculations in Asia, while the war in Afghanistan and India's rising power garner major American interest and concern. Moreover, opportunities for U.S. interests in Southeast Asia appeared limited at times.

The region recovered somewhat from the Asian economic crisis of 1997-98 but has been subject to persistent political instability and economic uncertainty that curbed U.S. and other foreign investment. The instability and uncertainty also sapped the political power and importance of ASEAN and its leading members, even though the organization more recently has been at the centre of growing Asian multilateral activism dealing with salient regional economic, political and security issues.

STRENGTHENING U.S.-SOUTHEAST ASIAN RELATIONS

U.S. attention to the region rose in the first decade of the twenty-first century. The U.S.-led global war on terrorism broadened and intensified U.S. involvement throughout Asia. Southeast Asia for a time became the so-called "second front" in the U.S. struggle against terrorism. The United States worked closely with its allies Australia, the Philippines and Thailand, as well as with others such as Singapore, Malaysia and Indonesia in various efforts to curb terrorist activities in Southeast Asia. Meanwhile, following the Asian economic crisis, a variety of Asian regional multilateral groupings centred on ASEAN and its Asian partners were formed and advanced significantly.

China's stature and influence in these groups and among ASEAN states grew rapidly amid burgeoning intra-Asian trade and investment networks involving China in a central role, and attentive and innovative Chinese diplomacy. China's rising prominence was seen by many to steer the region in directions that reduced American influence and worked against U.S. interests. On the other hand, the massive and effective U.S.-led relief effort in the wake of the 26 December 2004 Indian Ocean Tsunami disaster in South and Southeast Asia showed unsurpassed American power and influence and underlined the continuing importance of the United States for regional stability and well being.

Subsequent events saw Indonesia receive notable U.S. attention with former Secretary of State Rice and Secretary of Defence Donald Rumsfeld making separate trips during 2006 and President Bush making a visit in conjunction with the APEC summit in November. U.S. assistance to Indonesia in fiscal year 2006 was over US$500 million. Following the U.S. waiver in November 2005 of remaining legislative restrictions on military assistance to Indonesia, the Bush administration in March 2006 permitted sales of lethal military equipment on a case-by-case basis and the U.S.

Pacific Command in March endorsed "rapid, concerted infusion" of military assistance to the country.

Vietnam, the host of the November 2006 APEC leaders meeting, also received notable U.S. attention. The United States took the lead in negotiations on Vietnam's successful entry into the WTO. The United States was a close second to China as Vietnam's leading trading partner and it was the largest foreign investor in the country.

Secretary Rumsfeld visited Vietnam in 2005 and Vietnam modestly strengthened defence ties with the United States. U.S. military activism in the region also included Secretary Rumsfeld's and Secretary of Defence Robert Gates' participation at the annual Shangri-La defence forum in Singapore, and Secretary of State Colin Powell's regular attendance and Secretary Rice's less consistent attendance at the ARF; a strong programme of bilateral and multilateral exercises and exchanges between U.S. forces and friendly and allied Southeast Asian forces; an active U.S. International Military Education and Training (IMET) programme with regional governments; and the U.S. role as the top supplier of defence equipment to the leading ASEAN countries.

The U.S. military maintained what was seen as a "semi-continuous" presence in the Philippines to assist in dealing with terrorist threats, and it resumed in November 2005 a "strategic dialogue" with Thailand.

These moves added to progress made in U.S. security ties with Singapore (the largest regional purchaser of U.S. military equipment) and the upswing in U.S. military ties with Indonesia. The pace and scope of U.S. military activism in the region reinforced the tendency of U.S. allies and associates to find reliance on the United States and its regional defence structures preferable to reliance on other nascent but rising Asian regional security arrangements. Numerous bilateral and multilateral U.S. military exercises were valued by Asian partners. The United States promoted maritime security cooperation in the strategically important

Straits of Malacca by working with the states bordering the Straits, Singapore, Malaysia and Indonesia, to develop a command, control and communications infrastructure that will facilitate cooperation in maritime surveillance of the Straits.

Malabar, an annual U.S. exercise with India was broadened in 2007 to include forces from Japan, Australia and Singapore in a large naval exercise near the eastern entrance of the Malacca Strait. It involved over thirty warships including two U.S. and one Indian aircraft carrier battle groups.

U.S. foreign assistance to Southeast Asian countries increased along with substantial increases in the U.S. foreign assistance budgets prompted by the war on terrorism begun in 2001 and the Bush administration's Millennium Challenge Account (MCA) and Global HIV/AIDS Initiative (GHAI) begun in 2004. MCA rewards countries that demonstrate good governance, investment in health and education and sound free market policies.

GHAI is focused on dealing with the worldwide health emergency caused by HIV/AIDS. In addition to Indonesia, noted above, which received the bulk of US$400 million pledged by the U.S. government for relief from the December 2004 tsunami disaster in addition to a substantial U.S. aid programme, other large recipients of U.S. assistance included the Philippines, Vietnam, Cambodia and East Timor.

THE WAR ON TERRORISM IN SOUTHEAST ASIA

There was strong support in the United States for increased military and other counter-terrorism cooperation with Southeast Asian governments after the September 11, 2001 terrorist attacks on America. Nonetheless, several years of strong anti-terrorism efforts combined with the U.S.-led war in Iraq prompted Southeast Asian leaders to complain that the United States was conducting its fight against international terrorism in the wrong way. American actions were seen to radicalize Asia's Muslims and to cause growing domestic opposition to Southeast Asian governments friendly to the United States.

Meanwhile, Southeast Asian leaders also complained that U.S. preoccupation with the war in Iraq and the broader war on terrorism made U.S. leaders inattentive to Southeast Asia and towards other areas such as nation building, economic development and cooperation in an emerging array of regional multilateral organizations. The U.S. actions against terrorism in Southeast Asia were primarily bilateral, though some attention focused on regional organizations such as the ARF, APEC and ASEAN.

With U.S. encouragement, APEC members in particular undertook obligations to secure ports and airports, combat money laundering, secure shipping containers and tighten border controls. The U.S. military took the lead after 9/11 in arranging for close cooperation with the Philippines, a treaty ally of the United States, to deal with terrorists in the country. Joint and prolonged exercises allowed hundreds of U.S. forces to help train their Philippine counterparts to apprehend members of the terrorist group Abu Sayyaf, active in the southwestern islands of the Philippines. U.S. military supplies to the Philippines increased markedly; other U.S. military training activities helped to strengthen Philippine forces to deal with other terrorist and rebellious groups. These included members of the Jemaah Islamiyah reportedly active in training terrorists in the Philippines.

The United States awarded the Philippines the status of Major Non-NATO Ally (MNNA) in 2003, provided increased military assistance and welcomed the small contingent of Philippine troops in Iraq. The contingent was withdrawn in 2004 to save the life of a kidnapped Philippine hostage. The Bush administration also awarded increased military aid and MNNA status to Thailand, the other U.S. treaty ally in Southeast Asia, which sent nearly 500 troops to Iraq. Thailand also reversed its refusals in the 1990s and offered sites for the forward positioning of U.S. military supplies. Cobra Gold, the annual U.S.-led military exercise in Thailand, attracted participation from other Asian countries.

Meanwhile, Singapore developed a new security framework agreement with the United States involving counter-terrorism cooperation, efforts against proliferation of weapons of mass destruction (WMD) and joint military exercises. The United States also supported a Southeast Asian anti-terrorism centre in Malaysia. U.S. support for Indonesia focused at first on the Indonesian Police Counter-Terrorism Task Force and broader educational and other assistance designed to strengthen democratic governance in Indonesia in the face of terrorist threats. Restrictions on U.S. assistance to the Indonesian military were slowly eased, amid continued reservations in Congress over the military's long history of abuses.

Popular sentiment in Malaysia and Indonesia was strongly against a perceived bias in the Bush administration's focus against radical Islam as a target in the war on terrorism. This added to the prevailing unpopularity of the Bush administration on account of its policies in Iraq and towards the Palestinian-Israeli dispute. Nonetheless, the Bush administration worked hard to improve relations with Malaysia and particularly Indonesia, the world's most populous Muslim state whose government emphasized moderation and democratic values. U.S. aid, military contacts and high level exchanges grew as the Indonesian democratic administration made progress towards more effective governance.

THE UNITED STATES AND ASIAN MULTILATERALISM

Given the variety of regional groupings focused on ASEAN, U.S. policy choices in dealing with Asian multilateralism in Southeast Asia were complex. Elsewhere in Asia, prevailing circumstances did not call for significant

U.S. policy action on multilateral groupings. In Central Asia, the Shanghai Cooperation Organization (SCO) did not allow U.S. participation. In South Asia, the South Asian Association for Regional Cooperation (SAARC) saw the United States participate as an observer; in Northeast Asia, the United States already has been a leader at the Six-Party Talks; China and other participating powers were keen to transform this arrangement eventually into a more permanent regional group to address broader security concerns.

In the case of the Southeast Asian oriented groupings, U.S. policy debate revolved around a basic choice. Should the U.S. government continue with its recent level of activity in the region, or should it take steps to further advance the American profile in regional groupings? Spurred on by China's rising influence in Asian multilateral groups and with the strong encouragement of Japan, Australia, Singapore and others that the United States play a more prominent role in these groups, the U.S. government increased involvement in these organizations as it took a variety of steps to shore up U.S. relations with Southeast Asian governments.

The Bush administration by 2006 developed U.S. initiatives to individual Southeast Asian nations as well as to ASEAN and related regional multilateral groups. These initiatives were based on the U.S. position as the region's leading trading partner, foreign investor, aid donor and military partner. U.S. initiatives to ASEAN represented in part the Bush administration's efforts to catch up with ASEAN's free trade agreements and other formal arrangements with China, Japan and other powers.

Strong U.S. opposition to the military regime in Myanmar continued to complicate U.S. relations with ASEAN. That regime's crackdown on large demonstrations led by Buddhist monks in 2007 saw the Bush administration take the lead in pressing the UN and individual states to take actions against the repressive government. In contrast to other powers seeking closer ties with ASEAN, the United States did not agree to accede to the TAC, and remained ambivalent on participation in the East Asia Summit (EAS) meeting that required agreement to the TAC as a condition for participation.

U.S. officials also said they were not opposed to Asian regional organizations that excluded involved powers like the United States, but U.S. favour focused on regional groupings open to the United States and other concerned powers. The United States supported the ARF, the primary regional forum for security dialogue.

The Bush administration strongly supported APEC. President Bush at the APEC Summit Meeting in November 2006 urged APEC members to consider forming an Asia-Pacific Free Trade Area. The U.S. initiative was seen to underline U.S. interest in fostering trans-Pacific trade groupings in the face of Asian multilateral trade arrangements that excluded the United States. At the APEC Summit in 2007 President Bush proposed the formation of a democratic partnership among eight democratically elected members of APEC and India. President Bush in November 2005 began to use the annual APEC leaders' summit to engage in annual multilateral meetings with attending ASEAN leaders.

He and seven ASEAN leaders attending the APEC leaders meeting in 2005 launched the ASEANUS Enhanced Partnership involving a broad range of economic, political and security cooperation. In July 2006 Secretary Rice and her ten ASEAN counterparts signed a five year Plan of Action to implement the partnership.

In the important area of trade and investment, the U.S. and ASEAN ministers endorsed the Enterprise for ASEAN Initiative (EAI) launched by the U.S. government in 2002 that provided a road map to move from bilateral trade and investment framework agreements (TIFAs), which were consultative in relation to free trade agreements (FTAs) that are more binding. The U.S. FTAs were seen by U.S. officials as markedly more substantive, involving a variety of tangible commitments and compromises, than the FTAs proposed by China and other powers.

From these U.S. officials' perspective, such U.S. FTAs thus were harder to reach but had a much more significant effect on existing trade relations. The United States has bilateral TIFAs with several ASEAN states and in August 2006 the U.S. Special Trade Representative and her ASEAN counterparts agreed to work towards concluding an ASEAN-U.S. TIFA. The Bush administration followed its FTAs with Singapore and Australia with FTA negotiations with Thailand and Malaysia, but those negotiations stalled.

The initiation in 2005 of the U.S. presidential mini-summits with ASEAN leaders attending the annual APEC leaders meeting conveniently avoided the U.S.-ASEAN differences over Myanmar, which was not an APEC member. The Bush administration for a time accepted a working engagement with Myanmar as an ASEAN member at lower protocol levels.

Myanmar was represented in the ASEAN-U.S. Dialogue, and Secretary Rice shook hands with all ASEAN foreign ministers at the signing ceremony of the ASEANUS Enhanced Partnership in July 2006. Meanwhile, the Bush administration announced in August 2006 that it was planning to appoint an ambassador to ASEAN and that the Treasury Department intended to establish a financial representative post for Southeast Asia. Deputy Assistant Secretary of State for East Asian and Pacific Affairs, Scott Marciel, was appointed the U.S. ambassador to ASEAN in 2008.

The events examined above show clear direction in U.S. policy and interests in favour of greater American activism and involvement with Southeast Asia. The change in U.S. administration has strongly reinforced this trend. Of course, there remain serious complications and uncertainties. Heading the list is the ability and will of U.S. leaders to devote consistent attention to regional bilateral and multilateral relations in Southeast Asia, especially given the overriding U.S. policy preoccupations with the global economic crisis and major conflicts and confrontations in Southwest Asia, the Middle East and North Korea.

Meanwhile internal instability on the part of most leading ASEAN states could indicate that greater U.S. activism may result in few major accomplishments if ASEAN governments remain cautious about international commitments during a period of domestic political instability. Nonetheless, the positive publicity flowing from Secretary Clinton's trip to Indonesia, the planned visit of President Obama to the region in November 2009 to attend the APEC Summit and the new U.S. government's demonstration of greater interest in and flexibility towards relations with Southeast Asia suggest that enhanced U.S. activism, involvement and flexibility in the region may represent the most significant change in U.S. policy in Asia under an Obama administration.

The new U.S. government, when otherwise not challenged by major crises like the one recently generated by North Korea, seems inclined to adhere fairly closely to pragmatic and generally constructive U.S. policy approaches to key Asia issues followed in the later years of the George W. Bush administration.

7

India's Interests In The Indian Ocean

One of the key milestones in world history has been the rise to prominence of new and influential states in world affairs. The recent trajectories of China and India suggest strongly that these states will play a more powerful role in the world in the coming decades. One recent analysis, for example, judges that "the likely emergence of China and India ... as new global players-similar to the advent of a united Germany in the 19th century and a powerful United States in the early 20th century-will transform the geopolitical landscape, with impacts potentially as dramatic as those in the two previous centuries." India's rise, of course, has been heralded before-perhaps prematurely.

However, its ascent now seems assured in light of changes in India's economic and political mind-set, especially the advent of better economic policies and a diplomacy emphasizing realism. More fundamentally, India's continued economic rise also is favored by the scale and intensity of globalization in the contemporary world. India also is no longer geopolitically contained in South Asia, as it was in the Cold War, when its alignment with the Russia caused the United States and China, with the help of Pakistan, to contain India. Finally, the sea change in Indian-U.S. relations, especially since 9/11, has made it easier for India to enter into close political and security cooperation with America's friends and allies in the Asia-Pacific. Much of the literature on India has focused on its recent economic vitality, especially its highly successful knowledge-based industrial sector.

The nature and implications of India's strategic goals and behavior have received somewhat less attention. Those implications, however, will be felt globally-at the United Nations, in places as distant as Europe and Latin America, and within international economic institutions. It also will be manifest on the continent of Asia, from Afghanistan through Central Asia to Japan. Finally, and most of all, the rise of India will have consequences in the broad belt of nations from South Africa to Australia that constitute the Indian Ocean littoral and region. For India, this maritime and southward focus is not entirely new.

However, it has been increasing due to New Delhi's embrace of globalization and of the global marketplace, the advent of a new Indian self-confidence emphasizing security activism over continental self-defence, and the waning of the Pakistan problem as India's relative power has increased. Other, older, factors influencing this trend are similar to those that once conditioned British thinking about the defence of India: the natural protection afforded the subcontinent by the Himalayan mountain chain, and the problem confronting most would-be invaders of long lines of communications-the latter a factor that certainly impeded Japan's advance toward India in World War II.

The December 2004 tsunami that devastated many of the coasts of the Indian Ocean (IO) turned the world's attention to a geographic zone that New Delhi increasingly sees as critically important and strategically challenging. The publication of India's new Maritime Doctrine is quite explicit on the central status of the Indian Ocean in Indian strategic thought and on India's determination to constitute the most important influence in the region as a whole. The appearance of this official paper complements a variety of actions by India that underscore New Delhi's ambitions and intent in the region. Why does New Delhi care about the Indian Ocean region? India is, after all, a large nation, a subcontinent in itself. Why is it driven to exercise itself in a larger arena, one larger in fact than the South Asian subregion?

The reality is that while India is a "continental" power, it occupies a central position in the IO region, a fact that will exercise an increasingly profound influence on-indeed almost determine-India's security environment. Writing in the 1940s, K. M. Pannikar argued that "while to other countries the Indian Ocean is only one of the important oceanic areas, to India it is a vital sea. Her lifelines are concentrated in that area, her freedom is dependent on the freedom of that water surface.

No industrial development, no commercial growth, no stable political structure is possible for her unless her shores are protected." This was also emphasized in the most recent Annual Report of India's Defence Ministry, which noted that "India is strategically located vis-à-vis both continental Asia as well as the Indian Ocean Region."

From New Delhi's perspective, key security considerations include the accessibility of the Indian Ocean to the fleets of the world's most powerful states; the large Islamic populations on the shores of the ocean and in its hinterland; the oil wealth of the Persian Gulf; the proliferation of conventional military power and nuclear weapons among the region's states; the importance of key straits for India's maritime security; and the historical tendency of continental Asian peoples or powers (the Indo-Aryans, the Mongols, Russia) to spill periodically out of Inner Asia in the direction of the Indian Ocean.

The position of India in this environment has sometimes been compared to that of Italy in the Mediterranean, only on an immense scale. To this list may be added the general consideration that, in the words of India's navy chief, Indians "live in uncertain times and in a rough neighborhood. A scan of the littoral shows that, with the exception of a few countries, all others are afflicted with one or more of the ailments of poverty, backwardness, fundamentalism, terrorism or internal insurgency. A number of territorial and maritime disputes linger on. ... Most of the conflicts since the end of the Cold War have also taken place in or around the [Indian Ocean region]."

Confronted by this environment, India-like other states that are geographically large and also ambitious-believes that its security will be best guaranteed by enlarging its security perimeter and, specifically, achieving a position of influence in the larger region that encompasses the Indian Ocean. As one prominent American scholar recently noted, "Especially powerful states are strongly inclined to seek regional hegemony." Unsurprisingly, New Delhi regards the Indian Ocean as its backyard and deems it both natural and desirable that India function as, eventually, the leader and the predominant influence in this region-the world's only region and ocean named after a single state.

This is what the United States set out to do in North America and the Western Hemisphere at an early stage in America's "rise to power": "American foreign policy throughout the nineteenth century had one overarching goal: achieving hegemony in the Western Hemisphere." Similarly, in the expansive view of many Indians, India's security perimeter should extend from the Strait of Malacca to the Strait of Hormuz and from the coast of Africa to the western shores of Australia. For some Indians, the emphasis is on the northern Indian Ocean, but for others the realm includes even the "Indian Ocean" coast of Antarctica. In this same vein, one-probably not atypical-Indian scholar judges that "a rising India will aspire to become the regional hegemony of South Asia and the Indian Ocean Region, and an extra-regional power in the Middle East, Central Asia and Southeast Asia.

Ceteris paribus, a rising India will try to establish regional hegemony just like all the other rising powers have since Napoleonic times, with the long term goal of achieving great power status on an Asian and perhaps even global scale." India's strategic elite, moreover, in some ways regards the nation as the heir of the British Raj, the power and influence of which in the nineteenth century often extended to the distant shores of the Indian Ocean, the "British Lake." Writing about the hill station and summer capital of Simla in that period, historian James Morris has observed:

The world recognized that India was a great Power in itself. It was an Empire of its own, active and passive. Most of the bigger nations had their representatives at Simla, and the little hill station on the ridge cast its summer shadow wide. Its writ ran to the Red Sea one way, the frontiers of Siam on the other. Aden, Perim, Socotra, Burma, Somaliland were all governed from India. Indian currency was the legal tender of Zanzibar and British East Africa; Indian mints coined the dollars of Singapore and Hong Kong.

It was from Simla, in the summer time, that the British supervised the eastern half of their Empire. Upon the power and wealth of India depended the security of the eastern trade, of Australia and New Zealand, of the great commercial enterprises of the Far East. The strength of India, so many strategists thought, alone prevented Russia from spilling through the Himalayan passes into Southeast Asia, and the preoccupations of generals in Simla were important to the whole world. Historian Ashley Jackson is even more explicit in highlighting the Indian dimension in all of this. He writes that India under the Raj was a subimperial force autonomous of London whose weight was felt from the Swahili coast to the Persian Gulf and eastward to the Straits of Malacca.

There was, in fact, an "Empire of the Raj" until at least the First World War, in which Indian foreign policy interests were powerfully expressed and represented in the Gulf and on the Arabian and Swahili coasts, often in conflict with other British imperial interests. Perhaps unsurprisingly, this imperial "Indian" posture in the Indian Ocean reflects the strategic vision of many influential Indians today. A second motive for India, and one obviously related to the foregoing, stems from anxiety about the role, or potential role, of external powers in the Indian Ocean.

The late Prime Minister Jawaharlal Nehru summed up India's concerns in this regard: "History has shown that whatever power controls the Indian Ocean has, in the first instance, India's sea borne trade at her mercy and, in the second, India's very independence itself." This remains India's view. The Indian Maritime Doctrine asserts: "All major powers of this century will seek a toehold in the Indian Ocean Region. Thus, Japan, the EU, and China, and a reinvigorated Russia can be expected to show presence in these waters either independently or through politico-security arrangements." There is, moreover, "an increasing tendency of extra regional powers of military intervention in [IO] littoral countries to contain what they see as a conflict situation."

India's concern about external powers in the Indian Ocean mainly relates to China and the United States. The Sino-Indian relationship has improved since India's war with China in 1962 and the Indian prime minister's 1998 letter to the U.S. president justifying India's nuclear tests in terms of the Chinese "threat." Most recently, the Chinese premier paid a state visit to India in April 2005, during which the two sides agreed to, among various other steps, the establishment of a "Strategic and Cooperative Partnership for Peace and Prosperity." Chinese and Indian naval units also exercised together for the first time in November 2005.

However, and notwithstanding the probably episodic progress registered of late, China and India likely will remain long-term rivals, vying for the same strategic space in Asia. Beijing, according to former Indian external affairs minister Jaswant Singh, is the "principal variable in the calculus of Indian foreign and defence policy." In the words of one

Indian scholar, China's "rise will increasingly challenge Asian and global security. Just as India bore the brunt of the rise of international terrorism because of its geographical location, it will be frontally affected by the growing power of a next door ... empire practicing classical balance-of-power politics."

Another observer has recently judged that "there is no sign of China giving up its 'contain India' strategy which takes several forms: an unresolved territorial dispute; arms sales to and military alliances with 'India-wary countries' (Pakistan, Bangladesh, Burma and now Nepal); nuclear and missile proliferation in India's neighborhood (Pakistan, Iran and Saudi Arabia); and opposition to India's membership in global and regional organizations." Most recently, India's defence minister said in September 2005 that the Sino-Indian "situation has not improved. Massive preparations and deployments by China in the Tibetan and Sikkim border areas near Arunachal Pradesh and the Aksai Chin ... has created an alarming situation."

Narrowing its focus to the IO, India cannot help but be wary of the growing capability of China's navy and of Beijing's growing maritime presence. In the Bay of Bengal and Arabian Sea, especially, New Delhi is sensitive to a variety of Chinese naval or maritime activities that observers have characterized collectively as a "string of pearls" strategy or a "preparation of the battlefield." For Beijing, this process has entailed achieving the capability, and thereby the option, to deploy or station naval power in this region in the future. A key focus in this connection is Burma (Myanmar), where Chinese engineers and military personnel have long been engaged in airfield, road, railroad, pipeline, and port construction aimed at better connecting China with the Indian Ocean, both by sea and directly overland.

Some of this activity, moreover, spills over onto Burma's offshore islands, including St. Matthews, near the mouth of the Malacca Strait, and the Coco Islands (Indian until their transfer to Burma in the 1950s), in the Bay of Bengal. On the latter, China is suspected of maintaining a communications monitoring facility that collects intelligence on Indian naval operations and missile testing. In addition to this "presence" in Burma, China is pursuing a variety of infrastructure links with Southeast Asia through the Greater Mekong Sub-region program and is building container ports in Bangladesh at Chittagong, and in Sri Lanka at Hambantota-directly astride the main east-west shipping route across the Indian Ocean.

Elsewhere, and perhaps most ominously for India, China is constructing a large new naval base for Pakistan at Gwadar. India also remains somewhat nervous about the large U.S. military presence in the Indian Ocean to India's west-in the Arabian Sea and the Persian Gulf. India's Maritime Doctrine observes that "the unfolding events consequent to the war in Afghanistan has brought the threats emanating on our Western shores into sharper focus.

The growing U.S. and western presence and deployment of naval forces, the battle for oil dominance and its control in the littoral and hinterland ... are factors that are likely to have a long-term impact on the overall security environment in the [Indian Ocean region]."

In similar fashion, the 2004-2005 Annual Report of India's Defence Ministry states, "The Indian Navy maintained its personnel and equipment in a high state of combat preparedness due to the continued presence of multinational maritime forces in the Indian Ocean Region resulting in a fast pace of activities in the area."

On the other hand, the continuing development of ties with the United States lately seems to have moderated Indian sensitivity to the U.S. presence in the Arabian Sea. In September and October 2005, for example, the two sides conducted their first naval maneuvers-MALABAR 05-employing U.S. and Indian aircraft carriers, and this occurred in the Arabian Sea.

Many Indians, moreover, also recognize that because of Washington's desire to draw closer to India in response to overlapping "China" and "terrorism" concerns, the increased American role in the Indian Ocean region lately has increased India's "strategic space" and political-military relevance. Any decrease in the level of U.S. involvement in the region also would increase pressure here from China. Wariness about China also is a factor in recent Indian efforts to increase Japan's profile in the IO. This was most recently made manifest by the March 2005 Indo-Japanese agreement to develop jointly natural gas resources in the strategically sensitive Andaman Sea. In any case, as one retired Indian diplomat recently commented, "asking outside powers to stay away is a pipe dream."

Of particular note, this last realization has led New Delhi to discard its traditional rhetoric about the Indian Ocean as a "zone of peace." That language, along with "nonalignment" and a diplomatic approach marked by preachiness and a "moral" dimension, were the policies of an India that was weak. That India now belongs to history: "India has moved from its past emphasis on the power of the argument to a new stress on the argument of power." A third factor animating Indian interest in the Indian Ocean region is anxiety about the threat posed by Pakistan and, more broadly, Islam in a region that is home to much of the world's Muslim population. Formerly this may not have been an important consideration.

Today, however, Islamic civilization often finds itself at odds with the West and with largely Hindu India, and this conflict frequently will play out in the Indian Ocean region. India's Maritime Doctrine, for example, observed "the growing assertion of fundamentalist militancy fueled by jihadi fervor is factors that are likely to have a long-term impact on the overall security environment in the [Indian Ocean region]." In a similar vein, India's naval chief recently declared that the "epicenter of world terrorism lies in our [India's] immediate neighborhood." India, however, will approach these matters pragmatically, as illustrated by New Delhi's close ties with Iran.

A fourth motive for India in the Indian Ocean is energy. As the fourth-largest economy (in purchasing-power-parity terms) in the world, and one almost 70 percent dependent on foreign oil (the figure is expected to rise to 85 percent by 2020), India has an oil stake in the region that is significant and growing. Some Indian security analysts foresee energy security as India's primary strategic concern in the next twenty-five years and believe it must place itself on a virtual wartime footing to address it. India must protect its offshore oil and gas fields, ongoing deep-sea oil drilling projects in its vast exclusive economic zone, and an extensive infrastructure of shore and offshore oil and gas wells, pumping stations and telemetry posts, ports and pipeline grids, and refineries.

Additionally, Indian public and private-sector oil companies have invested several billion dollars in recent years in oil concessions in foreign countries, many of them in the region, including Sudan, Yemen, Iran, Iraq, and Burma. These investments are perceived to need military protection. The foregoing considerations are the primary ones for India in the region. However, there also are important commercial reasons for New Delhi to pursue a robust Indian Ocean strategy. In the Indian view, "the maritime arc from the Gulf through the Straits of Malacca to the Sea of Japan is the equivalent of the New Silk Route, and ... total trade on this arc is U.S. $1,800 billion."

In addition, large numbers of overseas Indians live in the region-3.5 million in the Gulf and Arab countries; they, and their remittances, constitute a factor in Indian security thinking. In light of these interests, India is pursuing a variety of policies aimed at improving its strategic situation and at ensuring that its fears in the theater are not realized. To these ends, New Delhi is forging a web of partnerships with certain littoral states and major external powers, according to India's foreign secretary, to increase Indian influence in the region, acquire "more strategic space" and "strategic autonomy," and create a safety cushion for itself. One observer states: "To spread its leverage, from Iran ... to Myanmar and Vietnam, India is mixing innovative diplomatic cocktails that blend trade agreements, direct investment, military exercises, aid funds, energy cooperation and infrastructure-building." In addition, India is developing more capable naval and air forces, and it is utilizing these forces increasingly to shape India's strategic environment.

THE U.S. RELATIONSHIP

India's pursuit of closer ties with its neighbors in the region and with key external actors in the region is not haphazard. Rather, and as one would expect, India is systematically targeting states that will bring India specific and tangible security and economic benefits. The relationship with the United States is intended to enhance and magnify India's own power, and it constitutes perhaps the most important measure that is intended, inter alia, to promote the realization of India's agenda in the Indian Ocean. The United States, of course, is the key external actor in the IO and has a more significant military presence there-in the Persian Gulf and Arabian Sea, Pakistan, east and northeast Africa, Singapore, and Diego Garcia-than it did even a few years ago.

Thus, America's raw power in the region has made it imperative that New Delhi, if it is to achieve its own regional goals, court the United States-at least for some time. The U.S. connection, of course, also promotes Indian goals unrelated to the Indian Ocean. This developing relationship has been abetted by common concerns about international terrorism, religious extremism, and the rise of China. It also is a fundamental departure from the past pattern of Indian foreign policy.

Since President William Clinton's visit to India in 2000 (the first visit by a president in decades) and, more recently, the realization by the George W. Bush administration of the importance of a rising India, as well as the 11 September 2001 terrorist attack on the United States, the two nations have embarked on a broad program of cooperation in a variety of fields, especially security. This cooperation has included Indian naval protection of U.S. shipping in the Malacca Strait in 2002, a close partnership in responding to the 2004 tsunami, combined military exercises, U.S. warship visits to India, a dialogue on missile defence, American approval of India's acquisition of Israeli-built Phalcon airborne warning and control systems, and an offer to sell India a variety of military hardware, including fighter aircraft and P-3 maritime patrol planes.

Indo-U.S. ties recently have advanced with particular speed. In March 2005, notably, an American government spokesman stated that Washington's "goal is to help India become a major world power in the 21st century. We understand fully the implications, including military implications, of that statement." This declaration was followed, in June 2005, by a bilateral accord, a ten-year "New Framework for the U.S.-India Defence Relationship," that strongly implies increasing levels of cooperation in defence trade, including coproduction of military equipment, cooperation on missile defence, the lifting of U.S. export controls on many sensitive military technologies, and joint monitoring and protection of critical sea lanes.

George Bush hosted a summit with Prime Minister Manmohan Singh in July 2005, promising to strive for full civil nuclear cooperation with India. In effect, the president recognized India as a de facto, if not de jure, nuclear-weapon state and placed New Delhi on the same platform as other nuclear-weapon states. India, reciprocating, agreed to assume the same responsibilities and practices as any other country with advanced nuclear technology.

These include separating military and civilian nuclear reactors and placing all civilian nuclear facilities under International Atomic Energy Agency safeguards;

implementing the Additional Protocol (which supplements the foregoing safeguards) with respect to civilian nuclear facilities; continuing India's unilateral moratorium on nuclear testing; working with the United States for the implementation of a multilateral Fissile Material Cut-Off Treaty; placing sensitive goods and technologies under export controls; and adhering to the Missile Technology Control Regime and to Nuclear Suppliers Group guidelines.

The American and Indian delegations also agreed to further measures to combat terrorism and deepen bilateral economic relations through greater trade, investment, and technology collaboration. The United States and India also signed a Science and Technology Framework Agreement and agreed to build closer ties in space exploration, satellite navigation, and other areas in the commercial space arena. Notwithstanding this dramatic advance in relations, which-assuming eventual congressional approval of implementing legislation-establishes a very close United States-India strategic relationship, some bilateral problems will persist. One is Pakistan.

The U.S. administration's policy now is to expand relations with both India and Pakistan but to do so along distinct tracks and in differentiated ways, one matching their respective geostrategic weights. From New Delhi's perspective, this is a distinct advance. Nonetheless, there will remain a residual Indian suspicion that any American efforts to assist Pakistan to become a successful state will represent means, potential or actual, of limiting Indian power in South Asia and the Indian Ocean. Such concerns have been diminishing; nonetheless, New Delhi will try to weaken or modify U.S. policies intended to strengthen United States-Pakistan ties, including continuing plans to sell the latter a large package of military equipment.

Other lingering problems in Indo-U.S. relations include New Delhi's close ties to Iran, apparently continuing Indian reservations about the large U.S. military presence in Southwest Asia and the Persian Gulf, India's pronounced emphasis on preserving its "strategic autonomy," and a persistent disinclination on India's part to ally itself with American purposes. In the latter regard, India, like China, Russia, and the European Union, will remain uncomfortable with a unipolar world and will do what it can to promote a multipolar order-in which it is one of the poles. New Delhi, therefore, will need to proceed adeptly to ensure that ties with the United States continue to develop and expand in such a way that its own policies and ambitions in the Indian Ocean are buttressed and advanced.

TOWARD THE ARABIAN SEAS

In addition to the U.S. relationship, New Delhi is seeking to increase India's profile almost omnidirectionally from India's shores. These efforts are intended to advance broad economic or security interests, including the "security" of the various "gates" to the Indian Ocean, and to cultivate ties with the nations adjacent to these choke points: the Strait of Hormuz (Iran), the Bab el Mandeb (Djibouti and Eritrea), the Cape of Good Hope and the Mozambique Channel (South Africa and Mozambique), and the Singapore and Malacca straits (Singapore and Thailand), among others. Certain Indian strategic and diplomatic initiatives also are aimed at gaining partners or client states once having strong ties with colonial or precolonial India.41

As noted above, India's Maritime Doctrine underscores the importance of the Arabian Sea region in the Indian view and highlights a growing attentiveness to challenges and opportunities arising there. Efforts by New Delhi to advance the Indian cause to its "near West" and in the "Arabian Seas" subregion have focused mainly on Pakistan, Iran, Israel, and several African states.

Indo-Pakistani relations have improved since early 2003, when Prime Minister Atal Bihari Vajpayee extended a "hand of friendship" to Pakistan; in January 2004, the two sides launched a peace process. India's aims in the current diplomatic interchange are to lessen the likelihood of an Indo-Pakistani military conflict, reduce pressure in Kashmir, and-especially-increase India's freedom to pursue great-power status and to maneuver elsewhere in South Asia, the region, and the world.

India does not expect an end, for a very long time at best, to difficulties in its relations with Pakistan. It is hoping, however, to manipulate the relationship in a manner that will leave India stronger and Pakistan weaker at the end of the day. As India is inherently the stronger party, any "closer" relationship between India and Pakistan will, in the long run, increase Indian leverage with respect to Pakistan and decrease Islamabad's ability to disregard Indian interests. As one Indian observer recently said, "India's long-term interest lies in changing Pakistan's behavior." The termination of support for perceived anti-Indian terrorism and more restraint in Islamabad's embrace of China, and eventually even the United States, are among India's goals.

Elsewhere in the Arabian Sea, India already has enjoyed considerable success in wooing Iran. That state, with its Islamic government, seems a strange partner for democratic India, but the two lands have long influenced each other in culture, language, and other fields, especially when the Mughals ruled India. India and Iran also shared a border until 1947. Iran sees India as a strong partner that will help Tehran avoid strategic isolation. In addition, economic cooperation with New Delhi (and Beijing) dovetails with Iran's own policy of shifting its oil and gas trade to the Asian region so as to reduce its market dependence on the West. For India, the relationship is part of a broader long-term effort, involving various diplomatic and other measures in Afghanistan and Central Asia, to encircle and contain Pakistan.

Obviously, New Delhi also regards the Iranian connection as helping with its own energy needs. Deepening ties have been reflected in the growth of trade and particularly in a January 2005 deal with the National Iranian Oil Company to import five million tons of liquefied gas annually for twenty-five years. An Indian company will get a 20 percent share in the development of Iran's biggest onshore oil field, Yadavaran, which is operated by China's state oil company, as well as 100 percent rights in the Juefeir oil field. India and Iran also have been cooperating on the North-South Transportation Corridor, a project to link Mumbai-via Bandar Abbas-with Europe.

There also is discussion of the development of a land corridor that would allow goods to move from India's Punjab through Pakistan, Iran, and Azerbaijan, then on to Europe. India and Iran also have been pursuing an ambitious project to build a 2,700-kilometer pipeline from Iran through Pakistan to India that would allow New Delhi to import liquefied natural gas. If finalized soon, the pipeline would be operational by 2010.

Security ties with Iran have been advancing as well. The parties have forged an accord that gives Iran some access to Indian military technology. There are reports-officially denied-that it also gives India access to Iranian military bases in the event of war with Pakistan. Other recent developments include the first Indo-Iranian combined naval exercises and an Indian effort to upgrade the Iranian port of Chahbahar, a move that could foreshadow its use eventually by the Indian Navy. This latter initiative presumably also responds to China's development, noted above, of a Pakistani port and naval base at Gwadar, a hundred miles east of Chahbahar.

The Indo-Iranian relationship is not without problems. Iran, of course, has never been happy about India's close ties with Israel. Most recently, Iran also was angered by a 24 September 2005 vote cast by India in support of an International Atomic Energy Agency (IAEA) resolution that potentially refers the Iranian nuclear weapons issue to the United Nations Security Council. The IAEA vote-passed despite one "no" vote and abstentions from Russia, China, and Pakistan, among others-follows several earlier hostile comments from India on the Iranian nuclear issue, including one calling on Tehran to "honor the obligations and agreements to which it is a party." The Indian vote was a blow to New Delhi's relations with Tehran.

However, while it may augur a more circumscribed future for this connection, it is more likely that the long-term effects of India's vote will be limited. The bilateral relationship is too important for both parties, and New Delhi and Tehran will do their best to ensure that ties remain on an even keel. India, however, recently has tried to reduce its vulnerabilities in the oil-rich but unstable Persian Gulf by moving beyond Iran and attempting to cultivate a broader and more diverse set of relationships there. The most significant recent development has been the new warmth in New Delhi's ties with Saudi Arabia, Iran's traditional foe in the Gulf and India's largest source of petroleum imports. Reflecting the change in the temper of Indo-Saudi ties, the new Saudi king was scheduled to be the main guest in New Delhi at the January 2006 Republic Day celebration.

This is a measure of the importance India attaches to its developing connection to Riyadh and an initiative undoubtedly noticed by the leadership in Iran. Moving farther westward, another key nexus is with Israel. While formal diplomatic ties date only from 1992, the two states have had important connections at least since the early 1980s. In recent years, numerous senior Israeli and Indian officials have exchanged visits, and military relations have become so close as to be tantamount to a military alliance. In 2003, following Pakistan's shoot-down of an "Indian" unmanned aircraft manufactured (and perhaps operated) by Israel, President Pervez Musharraf complained "that the cooperation between India and Israel not only relates to Pakistan, but the Middle East region as a whole."

Israel is now India's second-largest arms supplier after Russia, and India is Israel's largest defence market and second-largest Asian trading partner (after Japan). According to one estimate, India will purchase some fifteen billion dollars' worth of Israeli arms over the next few years. The two sides recently agreed to a combined air exercise pitting Israeli F-16s against Indian Su-30MKIs (an advanced derivative of the Soviet Su-27 Flanker). Israel possesses an Indian Ocean footprint that apparently encompasses the Bab-el-Mandeb, the southern entrance to the Red Sea and a key choke point, and probably points beyond. India's aim here is to link itself with another powerful state whose sphere thus intersects its own.

At the same time, New Delhi also seeks the advanced military equipment, training, and other help-probably including technology and advice on nuclear weapons and missiles-that Israel can sell or provide. The official publication of the Chinese Ministry of Foreign Affairs, World Affairs, claims that India is acquiring technology from Israel for its Agni-III missile as well as for a miniature nuclear warhead-which India would need were it to deploy a sea-based (i.e., Indian Ocean-based) strategic nuclear deterrent. Elsewhere in the western Indian Ocean, India forged its first military relationship with a Gulf state in 2002 when New Delhi and Oman agreed to hold regular combined exercises and cooperate in training and defence production. They also initiated a regular strategic dialogue and, in 2003, signed a defence cooperation pact.

The pact provides for the export and import of weapons, military training, and coordination of security-related issues. India and the Gulf Cooperation Council (GCC) also have signed a Framework Agreement for Economic Cooperation and have begun negotiations on a free trade pact. New Delhi's connections with Oman and the five other GCC states, however, still are relatively undeveloped. As one Indian observer noted recently, "With our growing dependence on imported oil and gas, stability in this region is crucial for our welfare and well-being. Around 3.7 million Indian nationals live in the six GCC countries.

They remit around $8 billion annually. ... The time has, perhaps, come for us to fashion a new and more proactive 'Look West' policy to deal with the challenges that we now face to our west." A month earlier, India's commerce minister offered the same view: "India has successfully pursued a 'look-east' policy to come closer to countries in Southeast Asia. We must similarly come closer to our western neighbors in the Gulf." Farther afield, India's ties with the states of Africa's Indian Ocean coast still are limited but are expanding.

Reminiscent of India's pre-colonial relationship with coastal Africa, New Delhi's key connections today are with some of the states in the Horn of Africa, South Africa, Tanzania, Mozambique, and especially the so-called African Islands, including Mauritius and the Seychelles. In the Horn, India is providing the force commander and the largest contingent of troops in the UN mission in Ethiopia and Eritrea. India also just concluded significant naval maneuvers in the Gulf of Aden, featuring drills with allied Task Force Horn of Africa units and a port call in Djibouti. At the other end of the continent, a noteworthy connection is developing with South Africa, through bilateral arrangements and a trilateral (India-Brazil-South Africa) relationship.

Developments in the security arena are striking and were underscored in late 2004 when the Indian Air Force conducted a combined air-defence exercise with its South African counterpart (and with participating American, German, and British elements)-the first combined air exercise ever conducted by India on the African continent. The participating Indian Mirage 2000 fighters deployed from north central India and flew-with help from newly acquired Il-78 aerial tankers-to South Africa via Mauritius. India and South Africa conducted combined naval drills off the African coast even more recently, in June 2005.

A visit by India's president to Tanzania in 2004 led to an agreement for increased training of Tanzanian military personnel in India and more frequent calls by Indian warships at Tanzanian ports. Farther south, Mozambique and India recently agreed to continue the joint patrols off the Mozambican coast begun during the African Union summit in Maputo in 2003. The governments also have begun to negotiate a defence agreement. New Delhi's links with the African Islands also are deepening. Since early 2003, India has been patrolling the exclusive economic zone of Mauritius, and it is negotiating a "comprehensive economic cooperation and partnership" agreement with what an Indian spokesperson calls this "gateway to the African continent."

In an April 2005 state visit, the Indian prime minister also reiterated India's commitment to "the defence, security and sovereignty of Mauritius." India also has initialed a memorandum of understanding with the Seychelles on defence cooperation: patrols of that nation's territorial waters, training of Seychelles military personnel, and-in early 2005-Indian donation of a patrol vessel to help with coastal defence. India, finally, has been very active in forging a close relationship with the Maldives, a connection undoubtedly reinforced by India's considerable material and other assistance in the aftermath of the December 2004 tsunami.

These island-nation initiatives were strengthened in September 2005 by the creation of a new defence ministry office headed by a two-star admiral charged with assisting such states. According to the Indian naval chief, these are "vital to India" and "friendly and well disposed," but their security remains fragile, and therefore India cannot afford to see any hostile or inimical power threaten them.

IN THE BAY OF BENGAL AND "FURTHER INDIA"

Complementing its westward orientation, India also has been diligent in cultivating closer relations with a variety of states in the Bay of Bengal and in Southeast Asia, often under the aegis of New Delhi's "Look East" policy. That approach, initiated in the early 1990s against the backdrop of a struggling Indian economy and the sudden disappearance of the Cold War framework, has been a stunning diplomatic success. As a consequence, India's ties with most of the states of the Bay of Bengal and Southeast Asia, except possibly Bangladesh, are better than they were only a few years ago. India has built a strong relationship with its immediate neighbor to the south, Sri Lanka. "India and Sri Lanka have forged new, close bonds. There is a new respect for India," according to one Sri Lankan observer.

This "respect," moreover, is sometimes reflected in reluctance in Colombo to challenge New Delhi, even on issues, such as the Sethusamudram Canal project, that could adversely affect important Sri Lankan interests. The Indo-Sri Lankan connection was solidified most recently by disaster relief in the aftermath of the 2004 tsunami, but a string of developments had already promoted close relations. A free trade agreement that came into force in 2000 has doubled bilateral commerce and increased significantly India's share of Sri Lanka's trade.

In addition, the two neighbors are moving steadily toward a Comprehensive Economic Partnership Agreement (CEPA). A defence cooperation agreement will soon expand Indian training programs for Sri Lankan troops, strengthen intelligence sharing, supply defence equipment (including transport helicopters) to Colombo, and refit a Sri Lankan warship. These states' first combined military exercise, EKSATH, took place in December 2004 and involved the Indian Coast Guard and Sri Lankan Navy. (New Delhi, however, has apparently rejected a Sri Lankan request for combined naval patrols against the Tamil "Sea Tigers.")

A memorandum provides for Indian help in reconstructing the vital Palaly airstrip on the Jaffna Peninsula in northern Sri Lanka. Colombo has rebuffed an Indian request that the field be reserved for use solely by Sri Lanka and India; however, taken in conjunction with a recent maritime surveillance pact, the accord could imply Indian utilization of that base eventually. New Delhi also has agreed to build a modern highway between Trincomalee, a Sinhalese pocket in the Tamil north and east, and Anuradhapura, in the Sinhalese heartland.

It will be named after former Indian Prime Minister Rajiv Gandhi, who was killed by a suicide bomber of the Liberation Tigers of Tamil Eelam (LTTE) in 1991. India likely also is contemplating the possibility of eventually using Trincomalee's legendary harbor. In a quiet deal in 2002, the Lanka Indian Oil Corporation, a wholly owned Indian government subsidiary in Sri Lanka, was granted a thirty-five-year lease of the China Bay tank farm at Trincomalee as part of its plan to develop petroleum storage there.

Also suggestive of wider Indian aims is the possible construction of a Trincomalee offshoot of the proposed pipeline between the southern Indian cities of Chennai and Madurai and Sri Lanka's capital, Colombo. Another of India's immediate neighbors is Bangladesh. The relationship has long been strained by such issues as illegal Bangladeshi migration, trade, and water use (notably New Delhi's "River-Linking Project"), but some improvement may be under way. Agreement by India and Bangladesh in January 2005 to move forward with an "Eastern Corridor Pipeline" to bring gas from Burmese fields through Bangladesh to India now appears to have been shelved.

Notwithstanding this setback, Indian Prime Minister Manmohan Singh visited Dhaka in conjunction with the thirteenth summit of the South Asian Association for Regional Cooperation in November 2005 and has invited Bangladeshi Prime Minister Kaleda Zia to visit India. One Bangladesh newspaper observed that "an improvement in the bilateral ties is seemingly an important foreign policy ... [goal] that New Delhi wishes to achieve. ... If India could put confidence building measures in place with Pakistan, its nuclear rival, we see no reason why Bangladesh's outstanding problems with India cannot be put behind." The "Look East" policy also has produced gains with the Association of Southeast Asian Nations (ASEAN). India became a sectoral partner in 1991, a full dialogue partner in 1995, and a member of the ASEAN Regional Forum in 1996.

In late 2004 India and ten ASEAN countries-meeting at the tenth summit in Vientiane-signed a historic pact for peace, progress, and shared prosperity. They also pledged to cooperate in fighting international terrorism and proliferation of weapons of mass destruction. The four-page accord and nine-page action plan envisage cooperation in multilateral fora, particularly the World Trade Organization; in addressing the challenges of economic, food, human, and energy security; and in boosting trade, investment, tourism, culture, sports, and people-to-people contacts. The pact commits India to creating a free trade area by 2011 with Brunei, Indonesia, Malaysia, Thailand, and Singapore, and by 2016 with the rest of ASEAN-the Philippines, Cambodia, Laos, Burma, and Vietnam.

Within ASEAN, India has focused particularly on developing close ties with Burma, Singapore, and, most recently, Thailand. Progress with Burma has been significant since New Delhi began to engage that nation about a decade ago, partly from concern about Chinese influence there. The emphasis now, however, is not mainly defensive but reflects India's regional ambitions, desire to use Rangoon from which to compete with China farther afield in Southeast Asia (including the South China Sea), and interest in Burmese energy resources, as well as its need to consolidate control in its own remote northeastern provinces.

Most recently, India's position in Burma was strengthened when strongman Khin Nyunt, known for pro-China inclinations, was deposed in October 2004 and placed under house arrest. Less than a week later, Than

Shwe, head of Burma's ruling military junta, visited India and signed three agreements, including a "Memorandum of Understanding on Cooperation in the Field of Non-Traditional Security Issues." The general also assured New Delhi that Burma would not permit its territory to be used by any hostile element to harm Indian interests. Soon thereafter, India and Burma launched coordinated military operations against Manipuri and Naga rebels along the frontier.

Indo-Burmese ties also are advanced by both countries' membership in the Bay of Bengal Initiative for Multi-Sectoral Technical and Economic Cooperation (BIMSTEC), the first setting in which two ASEAN members have come together with three countries in South Asia for economic cooperation. Significantly, neither China nor Pakistan is part of this grouping. These steps and others-resumption of arms shipments to Burma, New Delhi's acquisition of an equity stake in a natural gas field off Burma's coast, the proposed India-Burma Gas Pipeline, the reopening of the Indian and Burmese consulates in Mandalay and Kolkata, and a recent India-Burma naval exercise-all reflect a significant deepening in Indo-Burmese relations in recent years.

Burma ties as well into larger Indian agendas, to which eastward transportation is vital. New Delhi is building a road-the India-Myanmar-Thailand trilateral highway, a portion of the projected Asian Highway-connecting Calcutta via Burma with Bangkok. India also is building roads to connect Mizoram with Mandalay and has extended a fifty-six-million-dollar line of credit to Burma to modernize the Mandalay-Rangoon railroad. New Delhi is likely also to carry out port and transportation improvements at the mouth of the Kaladan River in western Burma, opening trade opportunities with Burma and Thailand and expanding access to India's northeast.

In addition, New Delhi has begun to study the feasibility of building a deep-water seaport at Dawei (Tavoy), on the Burmese coast, possibly allowing access from the Middle East, Europe, and Africa to East Asian markets without transiting the Malacca Straits. Taken together, these eastward transportation plans will give India an alternative route to the Malacca Straits sub-region as well as land access to the South China Sea. They reflect a land-sea strategy for projecting Indian influence to the east-a strategy intended to counter China's strategic ambitions in Southeast Asia and toward the Indian Ocean.

India's perceived need to compete with China in Southeast Asia, particularly in its littoral nations, has helped produce a courtship of Singapore. It also underscores the importance India attaches to key choke points-that it may need to block a Chinese move toward or into the Indian Ocean (the principal mission of the Indian bases in the Andaman and Nicobar Islands). Singapore is ideally situated to supplement the infrastructure in the Andaman; facilities there could, by the same token, allow India to project power into the South China Sea and against China. The Singapore relationship is modest but deepening.

Trade has been growing rapidly, surging by nearly 50 percent in 2004; a Comprehensive Economic Cooperation Agreement in June 2005 should boost trade further. In addition, a security pact in 2003 extended an existing program of combined naval exercises to encompass air and ground maneuvers and initiated a high-level security dialogue and intelligence exchange. Singapore and India held their first air exercise late in 2004 and their first ground exercises from February to April 2005, in India. Notably, in February and March 2005 their annual naval maneuvers took place for the first time in the South China Sea (vice "Indian" waters).

New Delhi also has stated willingness-in principle-to allow the Singapore Air Force to use Indian ranges on an extended basis. The developing Indian relationship with Thailand, finally, is a recent one and has been fed by, among other factors, Bangkok's growing concern with Islamic militants in Thailand's south: "The Thais know they are in a difficult situation and are looking left, right and center to see who is in the game on their side." A team of Indian intelligence officials visited Bangkok in November 2004; Thailand's National Security Council chief reciprocated the following month.

In addition, India's military has been coordinating closely with Thailand's navy and coast guard in and near the Malacca Strait, signing a memorandum of understanding in May 2005. Thailand also has been cooperating more than previously on matters related to the various insurgencies in India's northeast. More broadly, Bangkok welcomes the "rise of India," given Thailand's historical preference that no single power-not Britain or France in the nineteenth century, and not China today-achieve hegemony in its neighborhood. In any case, says one Thai pundit, "Our ancestors taught us to enjoy noodles as well as curry dishes." To this end, Bangkok is pursuing what it calls a "Look West" policy, and Thai officials have welcomed the Indian efforts to cultivate influence-potentially at China's expense-in Burma.

STRENGTHENING AND USING INDIA'S ARMED FORCES

Supplementing its diplomatic and political initiatives, India is shaping its growing military capability. These forces should be able, should the need arise, to: keep China's navy out of the Indian Ocean; enter the South China Sea and project military power directly against the Chinese homeland; project military power elsewhere in the Indian Ocean-at key choke points, on vital islands, around the littoral, and along key sea routes; and-in a presumably altered strategic environment-pose an important potential constraint on the ability of the U.S. Navy to operate in the IO.

At present, the overall thrust is to get weapons to project power, especially systems with greater lethality and reach.

To this end, India ordered $5.7 billion in weapons in 2004, overtaking China and Saudi Arabia and becoming the developing world's leading weapons buyer. Likewise, India stands as the developing world's biggest arms buyer for the eight-year period up to 2004.

The drive toward improved military capabilities is reflected in a variety of ongoing developments. The most significant development will be a strengthened nuclear-weapon strike capability relevant to the Indian Ocean as a whole. While land-based missiles may yet assume significance in this regard, New Delhi mainly is focused on equipping its navy and air force with nuclear capabilities that could be employed in a contingency.

India's intention to add a sea-based leg to its nuclear posture is longstanding and was a prominent feature of the Draft Nuclear Doctrine promulgated by India's National Security Advisory Board in 1999. The Cabinet Committee on Security also implicitly endorsed this goal in its 2003 restatement of many of the Doctrine's key points. Most recently, the new Indian Maritime Doctrine and the naval service chief, Admiral Arun Prakash, affirmed in September 2005 the importance of a sea-based leg.

INDIAN AIRPOWER

Another key development is the acquisition of an air force with longer range. A critical advance was the purchase in 2003 of Il-78 aerial tanker aircraft, New Delhi's first of the type. These tankers have supported the deployment of fighter and transport aircraft to a variety of far-flung locations, including South Africa and Alaska. Refueling also has recently allowed nonstop flights of Su-30s from Pune, their main operating base southeast of Mumbai, to Car Nicobar in the Bay of Bengal, a potential staging location adjacent to the Strait of Malacca and the South China Sea approaches to China's populous heartland.

A second airpower force multiplier will be the acquisition in 2007 of three Phalcon airborne warning and control system (AWACS) aircraft. These AWACS platforms, designed for 360-degree surveillance out to 350 nautical miles, will detect aerial threats and direct strike aircraft to targets.

Like the tankers, the AWACS will not have a mainly passive, defensive role; rather, they will allow other air assets to strike targets at greater distances and with much more effect. New Delhi also is developing an indigenous AWACS system, to be deployed by 2011. In addition, India's Tu-142M and Il-38 maritime surveillance/antisubmarine warfare aircraft all are receiving upgrades.

Finally, the Navy is raising three squadrons of Israeli-built Heron II unmanned aerial vehicles (UAVs) and probably will acquire P-3C Orions from the United States. India's air force also will achieve greater range and lethality with the acquisition of a variety of new combat aircraft-many of them clearly intended for strategic strike operations. In this regard, the planned acquisition of 190 long-range and air-refuelable Su-30 fighters (140 of which will be built from kits in India) through 2018 is particularly striking. New Delhi also has begun upgrading its fleet of Jaguar aircraft.

The package-an almost definitive sign that these aircraft will continue to have a nuclear strike mission-includes more modern navigation systems, new electronic countermeasures gear, and new armament pods. As these aircraft are capable of air-to-air refueling, the Il-78s significantly enhanced their radius of action. New Delhi also has ordered additional Jaguars (seventeen two-seat and twenty single-seats) from Hindustan Aeronautics Limited.

In addition, India plans to get 126 new multirole combat aircraft from a foreign supplier, either Lockheed Martin (the F-16), Boeing/McDonnell-Douglas (F-18 Hornet), Russia (MiG-35), Dassault Aviation of France, or Gripen of Sweden. Some of these airframes will be assembled in India. If Moscow and New Delhi can come to terms, at least four Tu-22M3s may be leased from Russia.

These Backfires have a range of almost seven thousand nautical miles and can carry a payload of about twenty-five tons-the equivalent of two dozen two-thousand-pound bombs, or a large number of standoff air-to-ground missiles.

India and Russia also are discussing the development and coproduction of a fifth-generation fighter aircraft. Many of these strike platforms will be equipped eventually with powerful, long-range cruise missiles. The joint Indo-Russian Brahmos, with a 290-kilometer range and supersonic speed, will be deployed first on Indian warships, but an aerial version is planned. As one observer comments, "India's co-development with Russia of the Brahmos missile for India's air (and naval) forces introduces ... a highly lethal, hybrid (cruise plus ballistic) missile that is most likely to be used as a conventional counterforce weapon against naval ships, ordnance storage facilities, sensitive military production facilities, aircraft hangars, military communication nodes and command and control centers."

A final aviation-related development, one reflecting the new over-the-horizon focus of the Indian Air Force, is the expected formation-with Israeli help-of an aerospace command that will feature a ground-based imagery center, intended to leverage India's growing space "footprint" for air force and missile targeting and battle space management. The new command will be linked to a military reconnaissance satellite system, expected to be operational by 2007.

INDIAN SEAPOWER

India's surface navy is to become more capable and lethal than today. India's first naval buildup occurred in the 1960s; there followed a period of robust growth in the mid-1980s. The latter expansion, marked by a focus on power projection, grew out of a perception of threat from the U.S. Navy, which was increasing its presence in the Indian Ocean. Prime Minister Indira Gandhi warned, "The

Ocean has brought conquerors to India in the past. Today we find it churning with danger." However, between 1988 and 1995 a retrenchment occurred, due to the disintegration of the USSR, a financial crisis in India (and East Asia), demands for social investment, and a virtually worldwide de-emphasis on military expenditures; the Indian Navy did not acquire a single principal surface combatant, either from abroad or from domestic shipyards.

The environment had changed again by the mid-1990s-as the international situation grew darker and the Indian economy strengthened-and the prospect is now for a navy that, if still modest in size, about forty principal combatants, will be significantly improved in quality. The surface navy currently consists primarily of a single vintage aircraft carrier, three new and five older destroyers, four new and seven vintage frigates, three new tank landing ships (LSTs), and assorted corvettes and patrol craft.

Within five years, this force likely will comprise instead two new (that is, to India) aircraft carriers, six new and only a few vintage destroyers, twelve new and a few older frigates, corvettes and patrol craft, five new LSTs, and a refurbished seventeen-thousand-ton ex-U.S. landing platform dock. All of the new warships, including the projected two aircraft carriers, will be much more formidable than their respective predecessors. For example, the Type 15A frigates now under construction in Mumbai will be equipped with sixteen vertical-launch Brahmos cruise missiles.

In addition, some warships are likely to be equipped eventually with U.S.-supplied Aegis radar systems. The carriers are particularly suited and intended for force projection. Moreover, with their aircraft and other weapons, they will constitute a quantum advance over the present carrier, INS Viraat, which is scheduled for decommissioning in 2010. One of the future carriers will be the 44,500-ton Soviet-built Admiral Gorshkov, now INS Vikramaditya, to be delivered in 2008.

The refitted ship will carry at least sixteen MiG-29Ks and six to eight Ka-31 antisubmarine and airborne-early-warning helicopters. India also has the option of acquiring, at current prices for up to five years, another thirty MiG-29Ks-a substantial increase in capability over the Harriers currently on the Viraat.

Also, Vikramaditya's range of nearly fourteen thousand nautical miles-vice the five thousand of Viraat-should represent a massive boost in reach. The other new aircraft carrier will be indigenously constructed, India's first; it was laid down in April 2005. The forty-thousand-ton vessel, designated an Air Defence Ship (ADS), is designed for a complement of fourteen to sixteen MiG-29K aircraft and around twenty utility, antisubmarine, and antisurface helicopters.

This will potentially equip the navy with two aircraft carriers by about 2010 (Vikramaditya and the ADS), thus allowing the service to maintain a strong presence along both the eastern and western shores. Indian naval leaders, however, envisage the navy as a three-carrier force-one on each coast and one in reserve-by 2015-20.

India continues to upgrade its existing submarine fleet while also developing or acquiring newer, more advanced boats. Many of these submarines are being fitted with cruise missiles with land-attack capabilities, reflecting the service's emphasis on littoral warfare. Over time, these cruise missiles almost certainly will be armed with nuclear warheads. The Indian Navy's principal subsurface combatants currently are four German Type 1500 and ten Russian-produced Kilo submarines. The Kilos are undergoing refits in Russia, including the addition of Klub cruise missiles, believed to have both antiship and land-attack capabilities at ranges up to two hundred kilometers.

The five boats already refitted with these weapons constitute the first Indian submerged missile launch capability. New Delhi is similarly upgrading one of its Type 1500s. The Indian government also recently authorized the purchase of six French-designed Scorpene submarines, with the option of acquiring four more. The first three boats will be conventional diesel-electric submarines, with subsequent ones incorporating air-independent propulsion. The design reportedly allows for the installation of a small nuclear reactor. The Scorpene contract apparently also provides for Indian acquisition of critical underwater missile-launch technology.

Other expected Indian submarine acquisitions include four to six Amur 1650 hunter-killer boats (SSKs) and two each of the more advanced versions of the Kilo and Shishumar submarines. India also has lately accorded higher priority to the construction of an indigenous nuclear-powered missile submarine, the Advanced Technology Vessel. Fabrication of the hull and integration (with Russia's assistance) of the nuclear reactor could already to be under way. In the long run, its main armament will be nuclear-armed cruise missiles.

Finally, New Delhi seems likely to lease from Russia two Akula II nuclear-powered attack submarines. Reportedly, Indian naval officers will begin training for these submarines at a newly built center near St. Petersburg in September 2005. These boats are normally configured with intermediate-range cruise missiles capable of mounting two-hundred-kiloton nuclear warheads, but India is expected instead to use the Brahmos cruise missile-eventually with a nuclear warhead-as their principal weapon.

BASING AND PRESENCE ASHORE

A better network of forward military bases is in prospect. One of the most important of its elements is INS (Indian Naval Station) Kadamba, a naval and naval air base-slated to be Asia's largest-under construction at Karwar (near Goa) on the Malabar Coast and recently inaugurated by Defence Minister Pranab Mukherjee. More centrally located with respect to the Indian Ocean than Mumbai, the site of India's longtime Arabian Sea naval complex, this facility will be India's first exclusive naval base (others are co-located with commercial and civilian ports). INS Kadamba will be able to receive India's new aircraft carriers; it is to become the home of several naval units beginning late in 2005 and, ultimately, of the headquarters of India's Western Naval Command.

It will reportedly serve as the principal base for the nuclear submarines that the Indian Navy is to lease from Russia and some that it will build indigenously. Farther south, India has been enhancing the infrastructure at Kochi (Cochin) in Kerala, where India's first full-fledged base for unmanned aerial vehicles recently was established.

The UAVs are providing the Navy a real-time view of the busy sea-lanes from the northern Arabian Sea to the Malacca Strait. As Kochi also is India's key center for antisubmarine warfare, the UAVs almost certainly also are employed for that purpose.

One observer, commenting on the strategic significance of this site, notes that "its situation, close to the southern tip of India's west coast and the central Indian Ocean, makes Cochin more than any other base a regional guard; a challenge to the United States in Diego Garcia [sic]; and the terminus of the trans-oceanic link with Antarctica."

In addition to Kochi, the Indian Navy is establishing UAV bases at Port Blair, the site of India's Andaman and Nicobar Command, and in the Lakshadweep Islands. The latter archipelago, off India's west coast in the Arabian Sea, is a key choke point between the Persian Gulf and the Malacca Strait that has until now received little attention from military planners.

New Delhi sees as even more strategically significant the Andaman and Nicobar Islands. It was to strengthen India's military presence in the Bay of Bengal that the unified Andaman and Nicobar Command were established in 2001. The islands had been recognized by the British as early as the 1780s as dominating one of the key gateways to the Indian Ocean. One analyst, writing from Port Blair, has claimed that "India was double-minded about retaining the islands until the 1998 Pokhran nuclear tests.

Top officials say the original plan was to abandon the Andaman and Nicobar Islands after exploiting its natural resources." India, for example, transferred the Coco Islands to Burma in 1954.

However, by 1962-in the aftermath of the war with China-New Delhi clearly was becoming sensitive to the archipelago's value, and in 1998 or before "the Vajpayee government woke up to the islands' huge strategic importance."

Whether or not India ever doubted its worth, the archipelago likely will have importance in the future-notwithstanding damage to infrastructure from the recent tsunami. India's navy chief has stated that "this theater will steadily gain importance ... in the coming years." Another Indian has characterized the new Andaman and Nicobar command as "India's ticket to strategic relevance" and "India's Diego Garcia."

In this connection, New Delhi almost certainly intends to use the islands as forward bases for cruise-missile-launching submarines, eventually with nuclear weapons. The islands also will play a key role in Indian efforts to parry Chinese inroads in Southeast Asia and to advance the "Look East" policy. Indian assistance in upgrading and developing the Iranian port of Chahbahar, the headquarters of Iran's third naval region, has been noted.

A construction initiative of another kind is the Sethusamudram project, also mentioned above, to cut through the Palk Strait and so permit Indian intercoastal shipping to avoid the long trip around Sri Lanka. Aside from its potential economic importance, such a route will enable warships from India's eastern and western fleets to quickly reinforce one another. In those terms the project is analogous to the 1914 completion of the inter-oceanic Panama Canal by the United States.

"MILITARY DIPLOMACY"

Supplementing the foregoing new weapons and military infrastructure advances, New Delhi also will use India's navy and air force, through "military diplomacy," to advance the Indian agenda in the Indian Ocean. India's new Maritime Doctrine declares, "Navies are characterized by the degree to which they can exercise presence, and the efficacy of a navy is determined by the ability of the political establishment of the state to harness this naval presence in the pursuit of larger national objectives." To this end, "the Indian maritime vision for the first quarter of the 21st century must look at the arc from the Persian Gulf to the Straits of Malacca as a legitimate area of interest."

India's navy and air force were indeed utilized in this manner in response to the December 2004 tsunami, perhaps the world's first global natural disaster. India was quick to extend help to Sri Lanka, the Maldives, and Indonesia. Indian relief operations were fully under way in Sri Lanka and the Maldives by day three of the tsunami (28 December), and the Indian military reached Indonesia by day four.

The subsequent relief operation was the largest ever mounted by New Delhi, involving approximately sixteen thousand troops, thirty-two naval ships, forty-one aircraft, several medical teams, and a mobile hospital.

Other recent instances of Indian military diplomacy include a continuing program of coordinated patrols with Indonesia in the Malacca Strait, naval surveillance of the Mauritius exclusive economic zone since mid-2003, and patrols off the African coast in connection with two international conferences in Maputo, Mozambique-the African Union summit in 2003 and the World Economic Forum conference the next year. An Indian Navy spokesman asserted that in these patrols the "Indian warships [were] demonstrating the Navy's emergence as a competent, confident, and operationally viable and regionally visible maritime power."

The Indian military also has been very active in pursuing combined exercises with a variety of IO partners. These maneuvers underscore the new flexibility and reach of Indian military forces. A Chinese newspaper, for example, commented that in one two-month period early in 2004 New Delhi conducted seven consecutive and quite effective combined exercises: "The scale, scope, subjects and goals of the exercises are unprecedented and have attracted extensive concern from the international community." That instance was not unique; the Indian Navy conducted simultaneous combined exercises with Singapore in the South China Sea and with France in the Arabian Sea in late February and early March 2005.

All this was followed immediately by a multiservice, combined planning exercise with the United Kingdom in Hyderabad; a naval exercise with South Africa and a port call by warships in Vietnam in June; and the deployment of a large flotilla to Southeast Asian waters in July. The agenda for late 2005 included naval maneuvers with the United States in the Arabian Sea in September, with Russia in the Bay of Bengal in October, and with France in the Gulf of Aden in November.

In addition, New Delhi partnered with Russia in a combined air-land exercise near the Pakistan border in October, and with the United States in November in a COPE INDIA air exercise (that latter in a location that clearly suggests mutual strategic concern about China). New Delhi, moreover, is expecting the advent of combined exercises with Japan's navy in the Sea of Japan and the Bay of Bengal in the not-too-distant future.

EXPECTATION OF INDIA IN THE INDIAN OCEAN

Over the past few years, India has placed itself on a path to achieve, potentially, the regional influence in the Indian Ocean to which it has aspired. To this end, New Delhi has raised its profile and strengthened its position in a variety of nations on the littoral, especially Iran, Sri Lanka, Burma, Singapore, Thailand, and most of the ocean's small island nations. India also has become a more palpable presence in key maritime zones, particularly the Bay of Bengal and the Andaman Sea. Of equal or greater importance, India's links with the most important external actors in the Indian Ocean-the United States, Japan, Israel, and France-also have been strengthened.

These are significant achievements, and they derive from India's growing economic clout and from a surer hand visible today in Indian diplomacy. Gaps inevitably remain in India's strategic posture. New Delhi will need to strengthen further its hand in coastal Africa and the Arabian Peninsula. More work also will be required to upgrade still somewhat distant relationships with Australia and Indonesia. At the same time, India will need to be more skillful than it has been in cultivating-or "compelling"-better relations with, and an environment more attuned to Indian interests in, Pakistan and Bangladesh.

Further, much will depend on the performance of the Indian economy and on India's ability to avoid domestic communal discord. Another variable will be the extent to which other states-particularly China and the United States but also Pakistan and others in southern Asia-are willing or able to offer serious resistance to India's ambitions. The future of political Islam is another wild card. However, barring a halt to globalization-one of the megatrends of the contemporary world-the rise of India in the IO is fairly certain. That will have a transforming effect in the Indian Ocean basin and eventually the world.

In the region, the rise of India will play a key role in the gradual integration of the various lands and peoples of this basin. Whether in the Arabian Sea or the Bay of Bengal, this trend-while still nascent-is already evident. The long-term result will be a more prosperous and globally more influential region. India's rise in the Indian Ocean also will have important implications for the West and China.

Perhaps most significantly, New Delhi's ascent suggests strongly that the ongoing reordering of the asymmetric relationship between the West and Asia will be centered as much in the Indian Ocean as in East Asia.

It was in the IO, moreover, that the effects of Western power first made themselves manifest in the centuries after 1500. On one hand, it would therefore not be surprising if it were here that the Western tide first receded. On the other, India's role will for a long time to come be no longer in opposition to the United States but in cooperation with it. Moreover, its rise will be welcomed by the United States and other "Western states" to the extent that it counteracts the challenge posed by China, the world's other salient rising power. Seen from Beijing, the rise of India in the Indian Ocean will be an opportunity but, even more, a challenge.

A strong and influential India will mean a more multi-polar world, and this is consistent with Chinese interests. Nonetheless, as China increasingly regards India-not Japan-as its main Asian rival, India's rise in the Indian Ocean also will be disturbing. As has been the case with virtually all great powers, an India that has consolidated power in its own region will be tempted to exercise power farther afield, including East Asia.

8

The Evolution Of Russian Strategy In The Indian Ocean

Perhaps the best single source for answering that question is the book written under government auspices by a distinguished group of Soviet military men and edited by Marshal V. D. Sokolovsky, *Military Strategy*. It is important to add, however, that the book's definition of peaceful coexistence *does not rule out national liberation struggles*, which in many Soviet-approved writings are called the only "legal and just wars." So in spite of the gains of WW II, *Military Strategy* would lead us to believe that today the Soviets consider war too dangerous to be an acceptable alternative strategy. But can we accept this as the definitive answer? Is Sokolovsky's book the final word and gospel?

Many have argued that the book was written for Western consumption and no more describes the official Soviet attitude toward war than *Catch 22* describes U.S. strategy. They can cite many instances of Soviet chicanery in the past that were effective to obfuscate her true intentions and can make a strong case that Sokolovsky's book is an attempt to satisfy all the disagreeing military and political factions in the RUSSIA and thus to be all things to all men.

On the other hand, Sokolovsky's book has been actively debated in the Soviet press, was quickly revised to meet Russian criticism, and is required reading for all Soviet officers. So it seems likely that *Military Strategy* does reflect the majority position of the Communist Party of the Soviet Union (CPSU) at the time of publication and revision but that it is not the only opinion of Soviet military and political leaders.

Since its publication the Soviets have devoted billions to achieve nuclear parity with the United States, a logical step if the book reflects official thinking and the Soviets feel, like the United States, that nuclear power deters war. But what about conventional forces—does not the Soviet conventional buildup in the late sixties cast doubt on the gospel according to Sokolovsky?

If all conventional wars lead to nuclear war and nuclear war is unacceptable and preventable by nuclear strength, then why a conventional buildup? The answer seems to be that after finally approaching nuclear parity with the United States in the 1967-69 period, the Soviets found themselves just as unhappy with the "launch or surrender" choice as the U.S. had become years earlier. Some Soviet military leaders had long argued for "flexible response," and it appears they finally won out. The Russia enters the seventies with her conventional forces in a transition state between a powerful but Eurasia-focused force to one with limited capability for employment away from Eurasia, when and where heavy opposition would not be anticipated.

It can be concluded, then, that *Military Strategy* probably does reflect the latest thinking of the political leaders of the RUSSIA: they do believe that nuclear missiles are the most important weapons in modern war and that nuclear war must be avoided, but growing Soviet international interests cause them to see also the need for strong conventional forces. Since a strategy is designed to reach a specific end, before we can identify the Soviets' more probable choice of strategies we must identify the ends she desires.

If we turn to the available literature and accept it literally, all we can find out is that she desires peace and prosperity for her people, for all the people of the world, and worldwide Communism. Pervading the entire list is the determination of the CPSU to remain in control of Russia's destiny. A threat to that control from any source could cause the goals to be drastically rearranged, and the avoidance of that eventuality would become the first priority.

FIRST STRIKE STRATEGY

The strategy of attaining first-strike capability and resultant world domination is not easily written off. The soviets enter the seventies with a lead over the United States of three hundred or so deployed or partially deployed land-based ICBM's. They trail the United States in long-range bombers and in the Polaris class of nuclear submarines, but their production of the latter ranges from four to twelve a year, either figure exceeding current U.S.-approved production schedules. They are also spending as much as the United States on military programs, excluding U.S. Vietnam costs, and they probably get more for the dollar.

Certainly, for example, they can maintain an armed force of approximately the same size as that of the U.S. for about $9 billion a year compared to $30 billion for the U.S. Their navy is already capable of interposing itself between the U.S. Navy and some future objective and is frequently noted in waters far from Soviet shores for the first time in its history (35 to 50 ships on station in the Mediterranean and token appearances in the Indian Ocean and even off U.S. waters). Her military fleet is heavily oriented toward missiles, causing some Navy authorities to argue that she has more firepower than the larger U.S. Navy.

Perhaps equally as significant is the Soviet Union's merchant fleet, which in 1969 ranked seventh in the world and which by 1975, if all contracts already let are delivered and if there are no U.S. countermoves, will exceed the U.S. merchant fleet in numbers and dry weight tonnage. Other indicators of her growing conventional military capability are her recent surge in modern aircraft production, including the world's fastest air-superiority aircraft, the Fox bat; the creation of a marine force; helicopter carriers for modern antisubmarine warfare (ASW) work; Red Berets, with unknown duties but looking suspiciously like the U.S. Green Berets; and extensive modernization and mechanization of her ground forces.

To this impressive list must be added her deployed, albeit not as yet fully operational, antiballistic missile (ABM) system around Moscow; her access to and possibility for exclusive use of Middle Eastern ports; and, most ominous of all, the deployment of more than two hundred powerful SS-9 ICBM's with the capability and possibility of modification as multiple independent re-entry vehicles (MIRV). It is the SS-9 and growing Polaris-type submarine fleet that most imply the possibility the Soviets might seek first-strike capability.

Is this conceivable? Brent Scowcroft has concluded that neither the United States nor the RUSSIA can achieve first-strike capability unless the other country actually cooperates. It does seem, however, that with the SS-9 plus MIRV's the RUSSIA could be reasonably confident of destroying the U.S. Minuteman force. But this would still not achieve first-strike capability because they would have to wipe out simultaneously the Polaris fleet, the U.S.-based B-52 fleet, some of the tactical aircraft in Europe, and the Asian B-52 fleet. They would also have to be confident that the United States, realizing its weaknesses and vulnerability, might not fire its missiles after receiving early warning and before impact.

In view of the Soviet military's past predilection for overwhelmingly favorable odds before acting and the conservativeness of the Politburo, this alternative must be rated as unlikely. Soviet decision-makers must further weigh the U.S. lead in MIRV development and consider the facts that the United States now plans to deploy an ABM net around its missiles and that by keeping a close watch on Soviet deployed SS-9s the U.S. can take several moves to counterbalance them if they approach the 400 to 450 number of launchers necessary to threaten Minuteman survival.

In light of these facts, it would be extremely difficult for the Soviet war and political planners to acquire confidence in the success of any first-strike attempt. Another important consideration in this area is suggested by Thomas Wolfe. He notes that the Soviets have previously learned what the United States can do when supposed or real gaps in military capability develop, the missile gap and moon race being the most conspicuous instances. When the Americans found or thought themselves to be behind, their superior technological base enabled them to close the gap quickly and even streak ahead. In such a race, the Soviets have learned, they lose. Why are the Soviets deploying SS-9s when smaller missiles could do the job of deterrence?

Henry Bradsher, a correspondent with years of duty in the RUSSIA, blames such Soviet inconsistencies on "momentum." He said that "it is a mistake to look for a rationale between Soviet military policy and their world policy." The fact is that once the Soviets plan a weapon system only failure can halt it. The explanation is not difficult to follow, since the plan is made by men chosen by the Politburo—the Politburo blesses the plan, and by fiat it is good, right, and perfect. No one dares argue, even months or years later, that the item is not needed, that it is redundant or uneconomical.

That would be admitting to fallacy in the central planning and, worse, the possibility of mistake by the planners. It is better in the RUSSIA to build it, test it, and deploy it. If it proves useless or ineffective, it might be scrapped or at least delayed for modification; if it works, the momentum of the system will insure its eventual production. Soviet warehouses are full of items that were produced even though they were not what were needed.

In 1967 a Soviet film entitled "Sovremenik" ("the modern man") dealt with just such a problem—its ending was realistic: an uneconomical and obsolete system under development, which the "modern man" tried to scrap in favor of a newer system, was nonetheless built, and the modern man was fired for obstructing the plan.

The Soviets would undoubtedly like to achieve first-strike capability, but in order to achieve it I believe that they would need the active cooperation of the United States, and they would be foolish to count on it. The evidence is that they seek nuclear equality, even numerical superiority, for the psychological advantage it gives them, but they still fear nuclear war more than any other single development. *Military Strategy* correctly describes the Soviet feeling that nuclear war is unacceptable. A first-strike strategy is therefore the least likely of their strategies as long as the United States maintains its defenses.

SPHERES OF INFLUENCE STRATEGY

The next least likely but desirable and possible strategy is the spheres strategy. Soviet history is full of instances of this approach to power politics, and it has attracted their military and political leaders. Stalin and Hitler once agreed on such a plan for Europe and the Middle East, and Churchill and Stalin discussed such a plan for Europe. Reports came out of European capitals in the late sixties that a worldwide division into spheres of influence is the Soviets' goal of the seventies. If this is their strategy, it helps explain why they worked so diligently in the late sixties to achieve superpower strength more favorably comparable to that of the United States.

The Soviet hope, it can be argued, is to convince the United States and the world that these two nations alone have the power to rule and control the world; that *de facto* spheres already existing in East and West Europe could be expanded to other areas, resulting in more world stability. The spheres approach fits Communist ideology and methodology. Ideologically, it divides the world into two camps, and methodologically it is efficient, establishing clear rules for the game of politics and enabling long-range and detailed plans to be formed and followed. It is, however, a utopian dream that reveals Soviet misunderstanding of Western values.

It is further made impossible by China's rising competition with the Russia in Asia, by Japan's and West Germany's booming economies, and the rising tide of nationalism in the world. No doubt the Russians would like a spheres-of-influence carving up of the world; but, as Castro, De Gaulle, Dubcek, and Mao have proved, spheres are easier to plan on paper than to put into effect and maintain. Practical Soviet foreign policy planners will have this fact driven home to them more and more each day as Chinese/Soviet relations deteriorate, East Europeans continue to resist domination, and the Arabs, under men like Nasser, continue to comply only partially with Soviet advice.

The Soviet leadership will see this and, as attractive as the idea may be, abandon any hopes for the success of a spheres strategy. The next two overall strategies that the Russia might select are, from her viewpoint, less desirable but more practical and probable. During the seventies she may well switch from one of these grand strategies to the other, then back and forth as the world situation and her immediate objectives change.

GENUINE DÉTENTE STRATEGY

This is not a likely strategy for the early seventies because it requires change—something the Soviets, like most governments, are hesitant and slow about doing. Genuine détente (instead of limited détente) refers to détente not only on the central issues but in the peripheral regions of the world also. In this strategy, instead of just accepting the status quo in Western Europe and avoiding tensions over Berlin, for example, the RUSSIA would restrict its activities in the Middle East, Asia, Latin America, and elsewhere. Economic and political competition would continue—the Soviets might well even continue to arm the North Vietnamese and Viet Cong, since that situation would have been under way before the strategy was adopted—but new conflict would generally be avoided.

Negotiations might be held on such topics as troop reductions in Europe, strategic arms limitation, the future return of islands to Japan, etc. Eastern Europe and the United States would be able to further develop trade and other so-called bridge-building moves. Why should the Soviets adopt such a strategy? China is probably the most important factor pushing the Soviets toward this choice. Soviet planners must consider—their actions bear this out—a two-front confrontation almost as bad as not having a nuclear deterrent. China's deliberate pugnacity and aggressiveness toward the RUSSIA threaten to create just such a two-front confrontation from the mid-seventies on.

The second reason for this strategy is Russia's economic problem. Her economy is slowing down perceptibly (from a 10–14 percent to a 4–6 percent growth rate in the sixties), and less military expenditures, especially in the strategic area, could help to improve this situation. The Russian consumer is expecting and demanding more. Also, since 1970 was the one hundredth anniversary of Lenin's birth, Soviet planners needed to present an abundance of consumer goods and a peaceful outlook to the populace as part of that celebration.

This is much easier to accomplish if the status quo is accepted, military expenditures drop, and the government is seriously involved in negotiations with the West to reduce tensions and the danger of war. Genuine détente, however, requires serious negotiations and an attitude of compromise and conciliation. The Russia has been notoriously lacking in these attributes. Nonetheless, her current improvement in military power (especially strategic) vis-à-vis the United States enables her to bargain, for a change, from an equal or near-equal position. A freezing of that situation could appear very advantageous to the Communists, especially considering the real technological gap that makes any race a harder and longer haul for them. Parity, too, might be counted on to convince many of the U.S. allies that they can no longer rely on U.S. protection and that the seventies would be a good time to accommodate with Russia.

This feeling could be encouraged by a more reasonable Soviet foreign policy. An eventual gain for Russia arising from this strategy might be a serious weakening of NATO and more Soviet trade and influence in West Europe. Thus genuine détente with the West in the seventies will have its Soviet supporters. Early indications of such a possibility are evident—the preliminary Strategic Arms Limitation Talks (SALT), the 1969 Warsaw Pact call for a European Security Conference, the Moscow Bonn agreements, the ever increasing East European–West European trade, Soviet calls for an Asian security pact, and Soviet conciliatory acts toward ancient Asian foes such as Nationalist China.

There are, however, factors which make this Soviet strategy unlikely to be adopted, the strongest being the resistance of the hardcore, still-Stalinist bureaucracy throughout the government, the party, and the Politburo. To many of them, genuine negotiation with the capitalists is anathema. When the motherland is in danger, concession might be acceptable, but not when Russia is so powerful and itself unthreatened. Certain military men and others should be expected to argue that a time of parity is a good time for international exploitation—to regain the revolutionary zeal of Communism and to use the developing Soviet conventional power to gain influence or control on the periphery of U.S. concern. Marshal Yepishev, Colonel Ribkin, and other Soviet military men representing this feeling frequently warn against trusting the West.

They claim that any negotiation acceptable to the West must be detrimental to the Communists and should be avoided. In the late sixties and early seventies several Soviet affairs experts, including Milovan Djilas, have written about the rising influence of the Soviet military. A careful analysis leads me to believe, however, that although military influence is indeed on the rise, as a result of circumstances like Red China and Czechoslovakia, there is no indication of a man on horseback in the wings.

The military is still led by men who have made a career out of following party dictates; and the CPSU, following the guidelines formulated by Trotsky and imposed by Stalin, can keep the military subordinate to its desires. In the seventies Soviet military men will on rare occasions speak out in muted terms against a political decision, but they will carry the day only when they convince a substantial number of the Central Committee or Politburo that the policy they advocate is for the good of the party. The military's probable resistance to a Genuine Détente Strategy will contribute to its rejection by politicians.

Though the process of globalization has not stopped, other regions are coming to play an increasing role, regions that only recently were objects, not subjects, of world policy. If before the center of the world economy was in Europe, and then a new pole emerged in North America, now the center of economic might lies in Asia. The globalizing world of the 21st century is now balanced on the backs of three whales. Europe's share in the Gross Domestic Product is about 20%, North America's is roughly 25%, while Asia's share is over 35% and rising fast.

The current financial crisis has accelerated the redistribution of economic might that was already underway. The United States is losing its role as the main locomotive of the global economy. Its political elite still hasn't had the resolve to put through the painful structural reforms necessary to normalize the economy. The growth of the GDP is due mainly to China and India which, right up until the industrial revolution, were the world's largest economies. Today the law of uneven development is clearly favoring the East, not the West. More than a third of world expenditures on Research & Development now come from Asia, in two decades or so Asian countries will have outpaced the United States and the European Union in this respect.

Economic shifts inevitably affect political influence and military might. However, existing global institutions created in the 20th century do not reflect the new realities. This could lead to a destabilization of the international system, to a worsening of tensions, and an increase in the conflict potential between "old" and "new" centers of power. In Asia there are a number of sources of tension as it is. An "arc of instability" exists on the continent; it includes zones of military conflicts prompted by internal and external factors. It is here that nuclear arms and ballistic missiles are proliferating.

The United States, with the help of NATO and other allies, is trying to play the role of "policeman" in the Eastern hemisphere. This strategy seems doomed to fail, while provoking an arms race in Asia. In addition to protracted wars in Iraq and Afghanistan, new armed conflicts may soon crop up (Iran, North Korea). For the multipolar world of the 21st century a new global security system is essential, one that reflects the new distribution of power in the international arena. This system must combine the imperatives of globalization with the tendencies of regionalization. In the modern world, along with sovereign states, regional associations are playing an increasing role.

These associations have different forms and functions, but all are turning into real subjects of international relations. For Russia, whose vital interests are inseparably tied to Europe and Asia, a Eurasian strategy has a special meaning. This strategy should overcome Russia's alienation from integrational processes in Europe and East Asia, while making the country a key link in the Eurasian sphere.

Such a strategy should avert a fragmentation of the modern world by connecting two of the three main centers of the global economy. Both Europe and East Asia maintain tight financial ties with the United States, but to a lesser degree with each other.

A Eurasian strategy would provide stability to the world economy's main "triangle", with Russia acting as an "economic bridge" between the East and the West. Our country possesses enormous transport, communications and energy potential for the creation of a new Silk Road. High-speed trains, turbo-conductors, main air routes and the Northern sea route will make transit through the Indian Ocean noncompetitive.

A Eurasian strategy will also allow Russia to form equal relations with neighboring countries in the West and the East, ones which currently have larger populations, GDPs and militaries. Such a strategy should make those countries partners interested in Russia's successful development, on which their own well-being will depend to some degree. Today Russia participates in any number of multilateral institutions and forums which represent both European and Asian states.

These include the OSCE, CIS (Commonwealth of Independent States), EEC (Eurasian Economic Community), CSTO (Collective Security Treaty Organization), SCO (Shanghai Cooperation Organization), SVMDA (Conference on Interaction and Trust Measures in Asia) and BRIC (Brazil, Russia, India and China). They differ in terms of composition, functions and effectiveness.

In public view ASEM (Asia-Europe Meeting), a forum for state leaders in Europe and Asia, has serious potential. ASEM accounts for more than half of the world's GDP and 60% of world trade. Today it comprises 45 countries, including all members of the European Union and ASEAN (Association of Southeast Asian Nations), as well as China, Japan and India. Russia and Australia are slated to join ASEM. ASEM's biannual summits focus on trade and economic ties. But in recent years the agenda has expanded to include security matters and global challenges. Moscow may now become one of the leaders of that forum.

It is hard to predict what the institutional infrastructure of Eurasia will look like several decades from now. But the need to create such an infrastructure is clear. Otherwise multipolar chaos may reign in a Eurasia fraught with the use of nuclear weapons. For Russia, strategic isolation is unacceptable. We must have reliable partners in Europe and Asia who take into consideration our legitimate interests. Unfortunately, at the government level, there is no strategy integrating European and Asian policy. We underestimated the approach of Kazakh President Nursultan Nazarbayev, which for years has shown the need to create a Eurasian union. Surrounded by gigantic neighbors it is understandable that Kazakhstan would strive for integration.

Nazarbayev's conception is based on "voluntary integration with equal rights, joint political and economic development of post-Soviet states, and the common promotion of CIS countries to strong positions in a global world". In my view, this allows one to arrive at the critical mass necessary for Russia, Kazakhstan and other members of a Eurasian union to successfully defend their interests in interactions with other key players in the Eurasian sphere.

The EEC customs union (including Russia, Belarus and Kazakhstan) is a first step in that direction. It is important to clearly define Moscow's objective interests and to propose parameters for a joint position that are acceptable to Ukraine, Kazakhstan and Belarus. One should recall that when NAFTA was created the United States made considered concessions to its neighbors as did Germany when the European Union was created. Russia should think in terms of its long-term interests, not short-term profits. To be sure, the littoral states would have taken note of the scrambling by NATO and India to deploy naval forces on a sea route that is crucial for the countries of the Asian region.

Trade and imports of oil by China pass through this sea lane. All the same, China has merely reported on the NATO deployment without any comments. Russia, on the other hand, didn't bother to report but preferred to swiftly respond. Even as the NATO naval force left for the Indian Ocean, it was stated in Moscow that a missile frigate from Russia's Baltic fleet - aptly named *Neustrashimy* [Fearless] - was already heading to the Indian Ocean "to fight piracy off Somalia's coast". Moscow claimed that the Somali government sought Russian assistance. Two days later, as the Indian Defence Ministry was making its announcement, it was revealed by the speaker of the Upper House of the Russian parliament, Segei Mironov, an influential politician close to the Kremlin, that Russia might resume its Soviet-era naval presence in Yemen.

Interestingly, Mironov made the announcement while on a visit to Sana, Yemen. He said Yemen sought Russia's help to fight piracy and possible terrorist threats and that a decision would be taken in Moscow to respond in accordance with the "new direction" of Russia's foreign and defence policies. "It is possible that the aspects of using Yemen ports not only for visits by Russia warships but also for more strategic goals will be considered," Mironov said. He further revealed that a visit by the president of Yemen, Ali Abdullah Saleh, to Moscow is scheduled in the near future and the issue of military-technical cooperation will be on the agenda. Significantly, Mironov explained that Yemen had threat perception regarding groups affiliated to al-Qaeda, which might be hiding in the Somalia region.

9

The Regional Response To The Superpowers' Naval Presence In Indian Ocean

The Indian Ocean has long been the hub of great power rivalry and the struggle for its domination has been a perennial feature of global politics. It is third-largest of the world's five oceans and straddles Asia in the north, Africa in the west, Indochina in the east, and Antarctica in the south. Home to four critical access waterways—the Suez Canal, Bab-el Mandeb, the Strait of Hormuz and the Strait of Malacca—the Indian Ocean connects the Middle East, Africa and East Asia with Europe and the Americas. Given its crucial geographical role, major powers have long vied with each other for its control, though it was only in the nineteenth century that Great Britain was able to enjoy an overwhelming dominance in the region.

With the decline in Britain's relative power and the emergence of two superpowers during the Cold War, the Indian Ocean region became another arena where the U.S. and the former Russia struggled to expand their power and influence. The U.S., however, has remained the most significant player in the region for the last several years. Given the rise of major economic powers in the Asia-Pacific that rely on energy imports to sustain their economic growth, the Indian Ocean region has assumed a new importance as various powers are once again vying for the control of the waves in this part of the world.

Nearly half of the world's seaborne trade is through the Indian Ocean, and approximately 20 percent of this trade consists of energy resources. It has also been estimated that around 40 percent of the world's offshore oil production comes from the Indian Ocean, while 65 percent of the world's oil and 35 percent of its gas reserves are found in the littoral states of this Ocean. Unlike the Pacific and Atlantic Oceans, almost three quarters of trade traversing through the Indian Ocean, primarily in the form of oil and gas, belongs to states external to the region. Free and uninterrupted flow of oil and goods through the ocean's SLOCs is deemed vital for the global economy and so all major states have a stake in a stable Indian Ocean region.

It is for this reason that, during the Cold War years when U.S.-Soviet rivalry was at its height, the states bordering the Indian Ocean sought to declare the region a "zone of peace" to allow for free trade and commerce across the lanes of the Indian Ocean. Today, the reliance is on the U.S. for the provision of a "collective good": a stable Indian Ocean region. This chapter examines the emerging Indian approach towards the Indian Ocean in the context of the country's rise as a major regional and global actor. This is an empirical analysis of India's role in the Indian Ocean region, not a theoretical exposition of the issue.

It argues that though India has historically viewed the Indian Ocean region as one in which it would like to establish its own predominance, its limited material capabilities have constrained its options. With the expansion, however, of India's economic and military capabilities, Indian ambitions vis-a-vis this region are soaring once again. India is also trying its best to respond to the challenge that growing Chinese capabilities in the Indian Ocean are posing to the region and beyond. Yet, preponderance in the Indian Ocean region, though much desired by the Indian strategic elites, remains an unrealistic aspiration for India, given the significant stakes that other major powers have in the region. In all likelihood, India will look towards cooperation with other major powers in the Indian Ocean region to preserve and enhance its strategic interests.

INDIA'S BACKYARD

As India's global economic and political profile has risen in recent years, it has also, not surprisingly, tried to define its strategic interests in increasingly expansive terms. Like other globalizing economies, India's economic growth is heavily reliant on the free flow of goods through the Indian Ocean SLOCs, especially as around 90 percent of India's trade is reliant on merchant shipping. Given India's growing reliance on imported sources of energy, any disruption in the Indian Ocean can have a potentially catastrophic impact for Indian economic and societal stability.

India's Exclusive Economic Zone in the Indian Ocean, that according to the Law of the Seas runs 200 nautical miles contiguous to its coastline and its islands, covers around 30 percent of the resource-abundant Indian Ocean Region. Any disruption in shipping across the important trade routes in the Indian Ocean, especially those passing through the "choke points" in the Strait of Hormuz, the Gulf of Aden, the Suez Canal and the Strait of Malacca, can lead to serious consequences for not only Indian but global economic prospects. Unhindered trade and shipping traffic flow is a sine qua non for the implementation of India's developmental process.

Non-traditional threats in the form of organized crime, piracy and transnational terrorist networks also make it imperative for India to exert its control in the region. Indian strategic thinkers have historically viewed the Indian Ocean as India's backyard and so have emphasized the need for India to play a greater role in underwriting its security and stability. India's strategic elites have often drawn inspiration from a quote attributed to Alfred Mahan: "Whoever controls the Indian Ocean dominates Asia. The ocean is the key to seven seas. In the 21st century, the destiny of the world will be decided on its waters."

This quote, though apparently fictitious, has been highly influential in shaping the way Indian naval thinkers have looked at the role of the Indian Ocean for Indian security. While sections of the Indian foreign policy establishment considered India the legatee of the British rule for providing peace and stability in the Indian Ocean, India's neighbours remain concerned about India's "hegemonistic" designs in the region.

Underlining the importance of the Indian Ocean for India, K.M. Pannikar, a diplomat-historian, called for the Indian Ocean to remain "truly Indian." He argued that "to other countries the Indian Ocean could only be one of the important oceanic areas, but to India it is a vital sea because its lifelines are concentrated in that area, its freedom is dependent on the freedom of that coastal surface." Pannikar was strongly in favour of Indian dominance of the Indian Ocean region much in the same mould as several British and Indian strategists viewed India's predominance of the Indian Ocean as virtually inevitable.

It has also been suggested that given the role of "status and symbolism" in Indian strategic thinking, India's purported greatness would be reason enough for Indian admirals to demand a powerful navy. In view of this intellectual consensus, it is surprising that India's civilian leadership was able to resist naval expansion in the early years after independence. India took its time after independence to accept its role as the pre-eminent maritime power in the Indian Ocean region and for long remained diffident about shouldering the responsibilities that come with such an acknowledgement.

The focus remained on Pakistan and China and the overarching continental mindset continued to dictate the defence priorities of the nation with some complaining that the Indian navy was being relegated to the background as the most neglected branch of the armed services. As the great powers got involved in the Indian Ocean during the Cold War years, India's ability to shape the developments in the region got further marginalized. India continued to lag behind in its ability to project power across the Indian Ocean through the early 1990s, primarily due to resource constraints and a lack of a definable strategy. It was rightly observed that "if the Indian Navy seriously contemplates power projection missions in the Indian Ocean, [the then Indian naval fleet] are inadequate ... it has neither the balance nor the required offensive punch to maintain zones of influence."

India, for its part, continued to demand, without much success, that "extra regional navies" should withdraw from the Indian Ocean, which met with hostility from the major powers and generated apprehensions in India's neighbourhood that India would like to dominate the strategic landscape of the Indian Ocean. India's larger non-aligned foreign policy posture also ensured that Indian maritime intentions remained shrouded in mystery for the rest of the world. It has only been since the late 1990s that India has started to reassert itself in the Indian Ocean and beyond.

This has been driven by various factors: the high rates of economic growth that India has enjoyed since the early 1990s have allowed the country to invest greater resources in naval expansion; the growing threat from non-state actors that has forced India to adopt a more pro-active naval posture; and, a growing realization that China is rapidly expanding its influence in the Indian Ocean region, something that many in the Indian strategic community feel would be detrimental to Indian interests in the long-term.

India has a pivotal position in the Indian Ocean because, unlike other nations in the region with blue-water capabilities, such as Australia and South Africa, India is at the centre and dominates the sea lanes of communication across the ocean in both directions. There are now signs that India is making a concerted attempt to enhance its capabilities to back up its aspiration to play an enhanced naval role in the Indian Ocean.

EXPANDING RESOURCE BASE

Sustained rates of high economic growth over the last decade have given India greater resources to devote to its defence requirements. In the initial years after independence in 1947, India's defence expenditure as a percentage of the GDP hovered around 1.8 percent. This changed with the 1962 war with China, in which India suffered a humiliating defeat due to its lack of defence preparedness and Indian defence expenditure came to stabilize at around 3 percent of the GDP for the next 25 years.

Over the past two decades, the military expenditure of India has been around 2.75 percent but since the country has been experiencing significantly higher rates of economic growth over the last decade compared to any other time in its history, the overall resources that it has been able to allocate to its defence needs have grown significantly.

The armed forces have long been asking for an allocation of 3 percent of the nation's GDP to defence. This has received broad political support in recent years. The Indian prime minister has been explicit about it, suggesting that "if our economy grows at about 8 percent per annum, it will not be difficult for [the Indian government] to allocate about 3 percent of GDP for national defence."

The Indian Parliament has also underlined the need to aim for the target of 3 percent of the GDP. India, with the world's fourth-largest military and one of the biggest defence budgets, has been in the midst of a huge defence modernization programme for nearly a decade that has seen billions of dollars spent on the latest high-tech military technology. This liberal spending on defence equipment has attracted the interest of Western industry and governments alike and is changing the scope of the global defence market.

As for the share of the three services, during the ten-year period between 1996-97 and 2005-06, the average share of the expenditure on the army, navy and air force was 57 percent, 15 percent and 24 percent respectively. Though the navy's share is the smallest, it has been gradually increasing over the years, whereas the share of other services has witnessed great fluctuations. The Indian navy saw its allocation go up by 10.5 percent and procurement spending rise by 17 percent in 2007.

In 2008-09, the navy's share of the total defence allocation was 18 percent, compared to 47 percent of the Army, and 53 percent of the Air Force. In the overall defence expenditure for the services, the ratio of revenue to capital expenditure is most significant in assessing how the services are utilizing their allocated resources.

Capital expenditure is the element that is directed towards building future capabilities. While the ratio of revenue to capital expenditure has been around 70:30 for the defence forces as a whole, there is huge variation among the services with the ratio of navy being 48:52.

Of the three services, it is the only one that is investing in future capabilities to a greater extent than current expenditure. Capital expenditure determines the trend of modernization and with 52 percent of its allocation going toward capital expenditure, the Indian navy is ahead of the other two services in its endeavour to modernize its operations.

Three key acquisitions by the Indian navy—long-range aircraft, aircraft carriers, and nuclear submarines—are intended to make India a formidable force in the Indian Ocean. While India's global aspirations are clearly visible in the modernization activities of the Indian navy, non-conventional threats to Indian and global securities have also risen in recent times, which might result in a change of priorities for the defence forces.

GROWING THREATS FROM NON-STATE ACTORS

Non-traditional threats to global security have grown exponentially and maritime terrorism, gun-running, drug trafficking and piracies are the major threats that India is facing from the sea-borders of the country. With vital shipping lanes passing through the area, India has been emphasizing the importance of maritime security in the Persian Gulf and the Gulf of Aden. Various terrorist organizations from Al Qaeda to Jammah Islamiah use maritime routes around India in the Indian Ocean region for narcotics and arms trafficking through which they finance their operations.

Indian intelligence agencies have warned the government that India might face seaborne attacks by terrorist groups against the nation's oil rigs, involving production and support platforms, along both coasts of India. Piracy in various parts of the Indian Ocean, such as the Malacca Straits and the Horn of Africa, is rampant, requiring a strong Indian maritime presence. In line with this perception, the Indian maritime doctrine states: "The Indian maritime vision for the twenty-first century must look at the arc from the Persian Gulf to the Straits of Malacca, as a legitimate area of interest." Most of the attacks and hijackings on the high seas are clustered in three areas: the Gulf of Aden and the eastern coast of Somalia; the coast of West Africa, particularly off Nigeria; and the Indonesian archipelago.

In the first quarter of 2008, more than half of all attacks took place in the Gulf of Aden. In 2008, at least 92 ships were attacked in and around the Gulf of Aden, more than triple the number in 2007 and an estimated $25 to $30 million was paid in ransom to Somali pirates. Following the hijacking off the coast of Somalia in September 2008 of the merchant vessel MT Stolt Valor, owned by a Japanese company with 18 Indian crew members on board, the Indian government authorized the Indian navy to begin patrols in the Gulf of Aden and escort Indian merchant vessels.

India has an economic interest in ensuring the protection of even non-Indian owned cargo ships in the Gulf of Aden shipping lanes, as around 85 percent of India's sea trade on the route is carried by foreign-owned ships, while around a third of India's total fleet of 900 cargo ships deployed in international waters are at risk. Patrolling by the Indian navy is intended to protect the nation's sea-borne trade and instil confidence in the sea-faring community as well as functioning to deter pirates. Russia, NATO and the EU forces have also started patrolling the region but efforts remain disjointed.

India has made a case that a peacekeeping force under a unified command is needed to provide security to international shipping in pirate-infested regions. In a first operation of its kind since the 1971 war with Pakistan, India's stealth frigate, INS Tabar, shot at and sank a pirate "mother vessel" in the Gulf of Aden, which later turned out to be a Thai trawler. Since the trawler was under the command of the pirates who refused to surrender, the Indian naval vessel fired in self-defence. This incident once again highlighted the Indian navy's capability on the high seas, witnessed earlier by the world in the conduct of tsunami relief operations and during the evacuation of Indian nationals in the Lebanon War of July-August 2006.

Moreover, the Indian navy asserted its autonomy and ability in the service of a collective good: the protection of global maritime trade. India used this act of its navy to project India as a country capable of protecting its maritime interests and commercial sea routes in international waters. While on the one hand the Indian navy demonstrated its might on the high seas, on the other, its ability to tackle terrorism in the homeland has come under scrutiny after terrorists managed to launch a severe assault on Mumbai in November 2008, hoodwinking the Indian navy and Coast Guard.

The terrorists succeeded in entering Mumbai by using a trawler, indicating a systemic failure of the Indian security agencies. It is the responsibility of India's Coast Guard to secure India's Exclusive Economic Zone, up to 200 nautical miles, whereas the blue water beyond is the navy's responsibility. Though dangers of terror attacks from the sea have long been apparent to Indian policymakers, no action was taken to strengthen the anti-terror defences. India's long coastline, with its inadequate policing, makes it easy to land arms and explosives at isolated spots along the coast. This was how explosives were smuggled into India in 1993 for the bomb blasts that crippled the Indian financial capital.

The same method was used again by the terrorists to attack Mumbai in 2008. The Indian Naval Chief took responsibility for inaction and underlined weak infrastructure for patrolling and surveillance of coastal areas. Despite clear intelligence inputs the Coast Guard and the navy failed to either spot or interdict the Pakistani ship that carried terrorists from an Indus creek near Karachi in Pakistan. It is clear that global threats from non-state actors are multiplying.

India will have to work with other major naval powers, not only to tackle problems such as piracy, but also to deal with the larger issues of security for sea-going commerce. Because the navy has proven itself adept at giving the Indian government sufficient leverage in operational situations in the Indian Ocean, its utility for India in projecting power and protecting its interests is only going to increase. Yet the biggest challenge to the Indian navy might come from the expansion of the prowess of that other Asian giant in the Indian Ocean: China.

CHINA'S FORAY IN THE INDIAN OCEAN

China emerged as the biggest military spender in the Asia-Pacific in 2006, overtaking Japan, and now has the fourth-largest defence expenditure in the world. The exact details about Chinese military expenditure remain contested, with estimates ranging from the official Chinese figure of $35 billion to the U.S. Defence Intelligence Agency's estimate of $80-115 billion. But the rapidly rising trend in Chinese military expenditure is fairly evident, with an increase of 195 percent over the decade 1997-2006. The official figures of the Chinese government do not include the cost of new weapons purchases, research or other big-ticket items for China's highly secretive military. From Washington to Tokyo, from Brussels to Canberra, calls are rising for China to be more open about the intentions behind this dramatic pace of spending increase and scope of its military capabilities.

The Chinese navy, according to the Defence White Paper of 2006, will be aiming at a "gradual extension of the strategic depth for offshore defensive operations and enhancing its capabilities in integrated maritime operations and nuclear counter-attacks." China's navy is now considered the third-largest in the world behind only the U.S. and Russia and superior to the Indian navy in both qualitative and quantitative terms. The Peoples' Liberation Army (PLA) Navy has traditionally been a coastal force and China has had a continental outlook to security. But with a rise in its economic might since the 1980s, Chinese interests have expanded and have acquired a maritime orientation with an intent to project power into the Indian Ocean.

China is investing far greater resources in the modernization of its armed forces in general and its navy in particular than India seems either willing to undertake or capable of sustaining at present. China's increasingly sophisticated submarine fleet could eventually be one of the world's largest and with a rapid accretion in its capabilities, including submarines, ballistic missiles and GPS-blocking technology, some are suggesting that China will increasingly have the capacity to challenge America. Senior Chinese officials have indicated that China would be ready to build an aircraft carrier by the end of the decade as it is seen as being indispensable to protecting Chinese interests in oceans.

Such an intent to develop carrier capability marks a shift away from devoting the bulk of the PLA's modernization drive to the goal of capturing Taiwan. With a rise in China's economic and political prowess, there has also been a commensurate growth in its profile in the Indian Ocean region. China is acquiring naval bases along the crucial choke points in the Indian Ocean not only to serve its economic interests but also to enhance its strategic presence in the region. China realizes that its maritime strength will give it the strategic leverage that it needs to emerge as the regional hegemony and a potential superpower and there is enough evidence to suggest that China is comprehensively building up its maritime power in all dimensions.

It is China's growing dependence on maritime space and resources that is reflected in the Chinese aspiration to expand its influence and to ultimately dominate the strategic environment of the Indian Ocean region. China's growing reliance on bases across the Indian Ocean region is a response to its perceived vulnerability, given the logistical constraints that it faces due to the distance of the Indian Ocean waters from its own area of operation. Yet, China is consolidating power over the South China Sea and the Indian Ocean with an eye on India, something that comes out clearly in a secret memorandum issued by the director of the General Logistic Department of the PLA: "We can no longer accept the Indian Ocean as only an ocean of the Indians.... We are taking armed conflicts in the region into account."

China has deployed its Jin class submarines at a submarine base near Sanya in the southern tip of Hainan Island in South China Sea, raising alarm in India as the base is merely 1200 nautical miles from the Malacca Strait and will be its closest access point to the Indian Ocean. The base also has an underground facility that can hide the movement of submarines, making them difficult to detect. The concentration of strategic naval forces at Sanya will further propel China towards a consolidation of its control over the surrounding Indian Ocean region. The presence of access tunnels on the mouth of the deep water base is particularly troubling for India as it will have strategic implications in the Indian Ocean region, allowing China to interdict shipping at the three crucial chokepoints in the Indian Ocean.

As the ability of China's navy to project power in the Indian Ocean region grows, India is likely to feel even more vulnerable despite enjoying distinct geographical advantages in the region. China's growing naval presence in and around the Indian Ocean region is troubling for India as it restricts India's freedom to manoeuvre in the region. Of particular note is what has been termed as China's "string of pearls" strategy that has significantly expanded China's strategic depth in India's backyard. This "string of pearls" strategy of bases and diplomatic ties include the Gwadar port in Pakistan, naval bases in Burma, electronic intelligence gathering facilities on islands in the Bay of Bengal, funding construction of a canal across the Kra Isthmus in Thailand, a military agreement with Cambodia and building up of forces in the South China Sea.

Some of three claims are exaggerated as has been the case with the Chinese naval presence in Burma. The Indian government, for example, had to concede in 2005 that reports of China turning Coco Islands in Burma into a naval base were incorrect and that there were indeed no naval bases in Burma. Yet the Chinese thrust into the Indian Ocean is gradually becoming more pronounced.

The Chinese may not have a naval base in Burma but they are involved in the upgradation of infrastructure in the Coco Islands and may be providing some limited technical assistance to Burma.

Given that almost 80 percent of China's oil passes through the Strait of Malacca, it is reluctant to rely on U.S.

naval power for unhindered access to energy and so has decided to build up its naval power at "choke points" along the sea routes from the Persian Gulf to the South China Sea. China is also courting other states in South Asia by building container ports in Bangladesh at Chittagong and in Sri Lanka at Hambantota as well as helping to build a naval base at Marao in the Maldives.

Consolidating its access to the Indian Ocean, China has signed an agreement with Sri Lanka to finance the development of the Hambantota Development Zone, which includes a container port, a bunker system and an oil refinery. The submarine base that China has built at Marao Island in the Maldives has the potential to challenge the U.S. navy in Diego Garcia, the hub of U.S. naval forces in the Indian Ocean.

It is possible that the construction of there ports and facilities around India's periphery by China can be explained away on purely economic and commercial grounds but for India this looks like a policy of containment by other means. China's diplomatic and military efforts in the Indian Ocean seem to exhibit a desire to project influence vis-a-vis competing powers in the region, such as the U.S. and India.

China's presence in the Bay of Bengal via roads and ports in Burma and in the Arabian Sea via the Chinese-built port of Gwadar in Pakistan has been a cause of concern for India. With access to crucial port facilities in Egypt, Iran and Pakistan, China is well poised to secure its interests in the region. China's involvement in the construction of the deep-sea port of Gwadar has attracted a lot of attention due to its strategic location, about 70 kilometres from the Iranian border and 400 kilometres east of the Strait of Hormuz, a major oil supply route.

It has been suggested that it will provide China with a "listening post" from where it can "monitor U.S. naval activity in the Persian Gulf, Indian activity in the Arabian Sea, and future U.S.-Indian maritime cooperation in the Indian Ocean." Though Pakistan's naval capabilities do not, on their own, pose any challenge to India, the combinations of Chinese and Pakistani naval forces can indeed be formidable for India to counter. The most far-reaching decision at the Budapest meet was NATO's decision to establish a naval presence in the Indian Ocean, ostensibly for protecting World Food Program ships carrying relief for famine-stricken Somalia.

Announcing the decision, a NATO spokesman said, "The United Nations asked for NATO's help to address this problem [piracy off Somalia's coast]. The ministers agreed that NATO should play a role. NATO will have its Standing Naval Maritime Group, which is composed of seven ships, in the region within two weeks."

He added that NATO would work with "all allies who have ships in the area now". Seven ships from NATO navies had already transited the Suez Canal on their way to the Indian Ocean.

En route, they will conduct a series of Persian Gulf port visits to countries neighboring Iran - Bahrain, Kuwait, Qatar and the United Arab Emirates, which are NATO's "partners" within the framework of the so-called Istanbul Cooperation Initiative.

The mission comprises ships from the U.S., Britain, Germany, Italy, Greece and Turkey. NATO's Supreme Allied Commander Europe, General John Craddock, acknowledged that the mission furthers the alliance's ambition to become a global political organization. He said, "The threat of piracy is real and growing in many parts of the world today, and this response is a good illustration of NATO's ability to adapt quickly to new security challenges."

Evidently, NATO has been carefully planning its Indian Ocean deployment. The speed with which it dispatched the ships betrays an element of haste, likely anticipating that some among the littoral states in the Indian Ocean region might contest such deployment by a Western military alliance. By acting with lightning speed and without publicity, NATO surely created a *fait accompli*.

STRING OF COINCIDENCES

By any reckoning, NATO's naval deployment in the Indian Ocean region is a historic move and a milestone in the alliance's transformation. Even at the height of the Cold War, the alliance didn't have a presence in the Indian Ocean. Such deployments almost always tend to be open-ended.

In retrospect, the first-ever visit by a NATO naval force in mid-September 2008 to the Indian Ocean was a full-dress rehearsal to this end. Brussels said at that time, "The aim of the mission is to demonstrate NATO's capability to uphold security and international law on the high seas and build links with regional navies."

In 2007, a NATO naval force visited Seychelles in the Indian Ocean and Somalia and conducted exercises in the Indian Ocean and then re-entered the Mediterranean via the Red Sea in end-September. The NATO deployment has already had some curious fallout.

In an interesting coincidence, on October 16, just as the NATO force was reaching the Persian Gulf, an Indian Defence Ministry spokesman announced in New Delhi, "The [Indian] government today approved deployment of an Indian naval warship in the Gulf of Aden to patrol the normal route followed by Indian-flagged ships during passage between Salalah in Oman and Aden in Yemen.

"The patrolling is commencing immediately." The timing seems deliberate. Media reports indicated that the government had been working on this decision for several months. Like NATO, Delhi also acted fast when the time came, and an Indian ship has already set sail.

Delhi initially briefed the media that the deployment came in the wake of an incident of Somali pirates hijacking a Japanese-owned merchant vessel on August 15, which had 18 Indians on board. But later, it backtracked and gave a broader connotation, saying, "However, the current decision to patrol African waters is not directly related [to the incident in August]." The Indian statement said, "The presence of an Indian navy warship in this area will be significant as the Gulf of Aden is a major strategic choke point in the Indian Ocean region and provides access to the Suez Canal through which a sizeable portion of India's trade flows."

Indian officials said the warship would work in cooperation with the Western navies deployed in the region and would be supplemented with a larger force if need and that it would be well equipped. But Delhi obfuscated the fact that the Western deployment will be under the NATO flag and any cooperation with the Western navies will involve the Western alliance.

Given the traditional Indian policy to steer clear of military blocs, Delhi is understandably sensitive. Clearly, the Indian warship will eventually have to work in tandem with the NATO naval force. This will be the first time that the Indian armed forces will be working shoulder-to-shoulder with NATO forces in actual operations in territorial or international waters. The operations hold the potential to shift India's ties with NATO to a qualitatively new level. The U.S. has been encouraging India to forge ties with NATO as well as play a bigger role in maritime security affairs.

The two countries have a bilateral protocol relating to cooperation in maritime security, which was signed in 2006. It says at the outset, "Consistent with their global strategic partnership and the new framework for their defence relationship, India and the United States committed themselves to comprehensive cooperation in ensuring a secure maritime domain. In doing so, they pledged to work together, and with other regional partners as necessary." The Indian Navy command has been raring to go in the direction of close partnership with the U.S. Navy in undertaking security responsibilities far beyond its territorial waters.

The two navies have instituted an annual large-scale annual exercise in the Indian Ocean - the Malabar exercises. This year's exercises are currently under way along India's western coast. The Indian Navy did inherit a lot from the Royal Navy in terms of force structuring, training, equipment, Operational philosophy and the maritime orientation itself. With the partition of India, the Indian Navy started its maritime saga with the World War vintage ships. Of course, UK continued to provide training, equipment and supported the growth of the Indian due to the colonial linkages as well as commercial considerations which obviously were very essential in rebuilding the economy of that country post world war.

Instead of listing the ships and their induction in a chronological manner, I propose to deal with the subject in a different manner to highlight events and acquisitions and then examine the strategic impact of such acquisitions and their relevance in the global context. Before we peg those events and relate them to the inductions, it would be necessary to understand the operating philosophy and the maritime doctrine that ought to serve as the foundation in force level structuring and planning. The maritime doctrine of course was conceptualized and released only very recently. It does not imply, though, that the in operated with out a doctrine or an operating philosophy.

It is just that the formalization of the Maritime Doctrine was long overdue. It also shows the coming of age of the in which did not hesitate to make the document public and allow a debate to develop on its understanding of the maritime scenario and its own role in shaping the maritime destiny of our nation. The in not only needed to understand its own role as an instrument of State Policy bat also was required to make a powerful statement of its growth, importance in the world arena and its operating philosophy.

MARITIME DOCTRINE

A careful analysis of the Maritime Doctrine indicates that by and large the IN has adopted the Mahanian approach to developing the force structure. It is quite different that there is no mention of this strategist in the

Maritime Doctrine of the Navy recently published. Admiral Thayer Mahan is acknowledged as the strategist behind the way that the U.S. Navy has developed. In accordance with the profound assertions of Mahan, the approach of the U.S. was to have the Carriers, Battle Ships and large ships for carrying the war to the enemy's shores and for power projection.

In addition, in our context, the Soviet influences indeed were inevitable with the induction and integration of vast numbers of Soviet origin ships, submarines, aircraft sensors and weapon systems. A large number of in personnel did train in the former Soviet Union, and it was very natural for some of the thinking to be influenced by the Soviet doctrines. Some of the Western Analysts believed that Indian Navy was just as bad or as good as the Soviet Navy! It was not before long that they were forced to change their opinion; more so, when they came in close contact with the Navy which was indeed a wonderful amalgamation of the West, the East and the indigenous in a manner to provide some very innovative solutions to the problems of integration of the best from both the worlds.

ACCELERATED GROWTH PROFILE

There were various value additions in the late 70s and 80s. Notably, these projects were designated as SNA, SNF, SNR, SNM etc, which signified the Soviet Aircraft, Destroyers, Missile ships, Mine sweepers respectively. Some also jokingly said that this actually linked up with the initials of Admiral Nanda. The formal transfer of the Maritime Reconnaissance role from the IAF in 1976 brought on additional responsibilities on the IN in terms of keeping the EEZ under surveillance in addition to meeting other strategic and tactical requirements. With the addition of the Super constellation aircraft INAS 312 and the commissioning of the IL38 squadron, the reach of the IN now covered the far corners of the Indian Ocean to make their presence felt.

This was also the period that the Nuclear Submarine Chakra was taken on lease in 1988 to learn the ropes as well as to send a message to our neighbours about the intentions of the country in exercising the strategic option as part of a triad at a later date. While the leased submarine was returned in 1991 and decommissioned later in Russia, Indian Naval personnel had been initiated to the intricacies of operation and maintenance of the nuclear asset. The addition of the Kamovs and modern Seakings for ASW and ASUW in the 90s provided those tactical options not available hitherto to the fleet commander.

The Sea Harriers provided punch to the fleet. The creation of the IMSF with the MARCOs added to the punch of the Service. This elite force provided options that were essentially needed with the specter of terrorism looming large in the sub continent. With the mindset of the cold war, western analysts linked the growth pattern to Indian hegemonistic designs in the IOR. Thus any new acquisition how ever relevant in the security context of this vast country was looked on with suspicion.

We had definitely learned to ignore these hollow drum beatings and full credit is due to the planners who vigorously pursued what was essential to address our national security concerns. With the breaking up of the Soviet Union, there was greater understanding of our role by the western nations. The long standing suspicion of our intentions and alignments slowly gave way to regular interactions and exchanges that lead to understanding of the inevitability of our presence and involvement in our immediate neighbourhood.

Regional exchanges such as the Milan in the late 90s conducted at Port Blair, exercises with Singapore, UK, France, U.S. and other maritime nations brought together the naval community and helped clear those doubts of our intentions. The most notable of such initiatives is definitely the International Fleet Review conducted at Mumbai in 2001 with whole hearted participation by over 28 Nations.

This was an effort to build the "bridges of friendship". This event itself has brought the Navy in to the big league and would pay rich dividends in the process of strategic alignments and in achieving a place of pre eminence in the IOR, Tsunami. Before I go in to some specific platforms and their significance, It is essential to touch up on the most recent event that has established our role in the IOR as a capable forward looking Maritime force willing to reach out and help in its neighbourhood.

The reference indeed is to the role played by the Indian Navy and the Indian Coast Guard soon after the Tsunami struck the region on 26 Dec 2005. Not only did the forces render help in our own affected areas, but also reached out to Indonesia, Sri Lanka and Maldives in a big way. Regrettably, the fact that thousands of marine troops were deployed with over 28 ships, many aircraft and Special Forces has not been given the kind of publicity that was given to the role of the western forces that reached these areas, on many occasions much later than our own forces.

But the Nations which received such help from the Navy and the Coast Guard are well aware of the capabilities of these two forces in the region. This automatically elevates our standing in the region and would propel us to greater regional leadership role in the 'Asian Century'. Let us now look at some of the vital components of the Navy that have always raised doubts even amongst our own military planners. The ensuing discussions should answer some of the FAQs. While the Navy has believed in the concept of the fleet with the Carrier at the center, except for very brief period, we have not been able to have at least one available at all times for our missions.

It is for this reason, that Successive Chiefs of Naval Staff have insisted that we need a minimum of three carriers to cater for one of them not being available due to refit schedules etc. The induction of Viraat again a second hand Carrier was to be a short term solution. After much haggling over price and other related issues, it appears that the

Russian Carrier Gorshkov to be named as Vikramaditya will at best be a replacement for Viraat.

The Gorshkov Package has this Carrier being refitted in the Russian yard to be able to operate the Mig29 aircraft to provide the air defence and strike options for the Indian Fleet. The ADS the indigenous Carrier to be built in Kochi would provide the Nation and the Navy for the first time with a real challenge and a capability in building a war vessel of this size this time to meet our security demands in to the next century. It is still to be seen how the interface between the LCA and the ADS as well as Vikramaditya would be achieved.

STRATEGIC ASPECTS

The naval planners were acutely aware of our limitations in terms of a strategic deterrent. While the ATV project was to look at the nuclear option for our sub surface forces, it was decided to get a nuclear submarine on lease. The intention was definitely to learn more about the aspects of maintenance and operation of a nuclear sub in our waters. In this case though the submarine was not carrying nuclear weapons, it demanded full attention and creation of technical special to type facilities as it was nuclear propelled.

Unfortunately, it has taken a long time for us to provide the sea based deterrence that is so crucial to our declared nuclear posturing. There are reports, that two Akula type submarines would be leased shortly for meeting the interim strategic requirements. Hopefully, this should provide those interim strategic solutions that the nation is looking at.

THE FUTURE

What do we see as we look at the future? We do have a professional Navy that has come of age and is ready to take on the responsibility in the IOR. While the addition of Vikramaditya, ADS, Mig 29s, LCA, the Akulas and stealth Ships are certain, the induction of P3C Orions, E3C Hawk Eye and high tech equipment to enable net centric warfare is definitely on the cards. The Navy also seems to have come to grips with the challenges of managing the quality of its manpower by pro active policies and HRD initiatives.

The commissioning of world class training academy at Ezhimala, the alternate base at Karwar and augmenting of the infrastructural facilities would only strengthen the nations resolve to invest in this potent instrument of national will.

The oceans are central to our trade and economy initiatives in the coming century as well. The fact that the nation has to depend on the seas for trade transactions of over 97% itself emphasizes the need for a strong Navy that can protect the shipping lanes and thwart any threats to the vital sea trade.

The energy security is interlinked with the development of the nation's economy as never before. The growing energy demands can only be met by huge imports to our ports. Even today our share of domestic production of oil is just about 30% leaving the balance to be imported. As one traces the origin and development of the Indian Navy, there is a sense of accomplishment in what has been done. It is not that the sailing has been smooth through out its passage of history.

The vision of the leaders and the strength of the policies have enabled the Indian Navy to assume the mantle of leadership in the region. India can not be oblivious to the fact that it still has to develop its other maritime constituents to be recognized as a maritime power. Thus the simultaneous development of our Ports, Infrastructure, fishing, Scientific and Technological ability to harness the oceans assumes as much or even more importance in the present day context.

The Navy is not an end by itself but a means to fulfill national aspirations in the century of the seas. The Navy should continue to provide the security umbrella and also provide leadership role in the IOR with its potential and demonstrated strengths.

MULTINATIONAL NAVAL EXERCISES

China's forays in the Indian Ocean date back to 1985 when the PLAN made port calls to South Asian ports in Pakistan, Bangladesh and Sri Lanka. Pakistan emerged as an important partner in South Asia for China and today their cooperation covers a wide spectrum of political, economic and strategic issues including the sale and joint development of military hardware and nuclear cooperation. Both sides have also engaged in bilateral/multilateral naval exercises. Commenting on the first ever joint exercise with the Pakistani Navy held off the coast of Shanghai in 2003, Rear Admiral Xiu Ji, a Chinese navy official observed that the exercises were 'the first [for China] with any foreign country'.

Two years later, the second bilateral exercise was held in the Arabian Sea in November 2005. In 2007, Pakistan hosted a multinational naval exercise, Aman 2007, off Karachi and invited the PLAN to join the exercises. Beside the Pakistani Navy ships, warships from Bangladesh, China, France, Italy, Malaysia, the United Kingdom, and the United States engaged in maneuvers in the Arabian Sea.

Interestingly, the Commander of the Chinese flotilla Luo Xianlin was designated as the tactical commander for the joint maritime rescue exercise and the PLAN missile frigate 'Lianyungang' was entrusted with the coordination of the exercise. The exercises were significant since it provided the PLAN with the opportunity to coordinate complex maneuvers with other naval forces. In 2009, the PLAN once again participated in Aman 2009, which was held in the Arabian Sea, and this time it carried out exercises along with 19 foreign naval ships.

Although the PLAN has engaged in bilateral and multinational naval exercises, it is important to point out that deployments for multinational operations are relatively different and more complex. Conducting multinational operations involves structured communication procedures, synergy among different operational doctrines, establishing mutually agreed rules of engagement (RoE), helicopter controlling actions, and common search and rescue procedures, which the PLAN is still developing.

SHIFTING GEOGRAPHY OF PEACE MISSION

A close partnership between China and Russia is evident in the maritime domain and rests on joint naval exercises, Chinese acquisition of Russian naval hardware including ships, submarines and aircraft and high-level naval exchanges. In 1999, the two navies conducted a joint naval exercise that involved the Russian Pacific Fleet and the PLAN's Eastern Fleet and the 2001 joint exercises included Russian strategic bombers. Peace Mission 2005, another naval exercise involving the PLA Navy and the Russian Navy was conducted under the Shanghai Cooperation Organization (SCO), the six-nation security group.

The exercises were conducted off the East Russian coast-Shandong Peninsula in northeastern China. Peace Mission 2007 focused on counter-terrorism and was conducted on land. Interestingly, the two sides utilized their presence in the Gulf of Aden and conducted Blue Peace Shield 2009, a joint exercise involving counter piracy operations, replenishment-at-sea, and live firing. The exercise showcased Chinese intention to be more transparent in its deployment, test interoperability with foreign navies and the PLAN's ability to engage in a range of operations in distant waters.

ENGAGING STRAITS OF MALACCA LITTORALS

China has adopted diplomacy as a tool to ally apprehensions among the Straits of Malacca littorals thus setting aside their fears that Beijing may deploy its navy in times of crisis to escort Chinese flagged vessels transiting through the Strait. Further, China is averse to any extra regional attempts to deploy naval vessels in the Strait for the safety of merchant traffic transiting. For instance, in 2000, it strongly objected to Japanese attempts to deploy vessels to patrol the Straits of Malacca where shipping had been threatened by piracy.

Instead, it has offered financial and technological assistance to improve the safety and security of merchant traffic transiting the Strait of Malacca. In 2005, during the International Maritime Organization (IMO) meeting in Jakarta, China reiterated its position of supporting the littoral states in enhancing safety and security in the Strait. In 2005, China offered to finance the project for the replacement of navigational aids damaged during the 2004 Indian Ocean Tsunami and the estimated cost for the project is pegged at $276,000.

BENEFITS OF MULTINATIONAL EXERCISES FOR PLAN

Multinational naval operations are fast gaining higher priority in the PLAN's strategic thinking. There are at least three reasons. The first relates to the 2004 Indian Ocean Tsunami and the international disaster relief operations in Southeast Asia-South Asia. PLAN's conspicuous absence in the operations had exposed the limitation of a rising power and its navy. As a result, China was excluded from the core group comprising the United States, Australia and India who quickly deployed their ships for relief efforts.

The Chinese Navy's absence might also be attributed to its lack of experience in working with multinational forces. The second reason for participation in multinational exercises is prospects for interoperability with international navies. Further, these operations assist the PLAN in identifying international trends in naval weaponry, gathering information on operating procedures and gaining a better understanding of the changing nature of naval warfare.

The third reason is that multinational exercises help China showcase to the international naval community its military industrial prowess and PLAN technological sophistication. Yet, China embraces selective maritime multilateralism. For instance, China did not participate in the U.S. Naval War College's International Sea Power Symposium in Newport. This year's event is the 40th anniversary and provides an occasion for the heads of the world's navies and coast guards to discuss issues of mutual interest. The 2009 Symposium focused on common maritime challenges and explored prospects for enhancing maritime security cooperation, including combating piracy.

IMPEDIMENTS TO CHINESE MARITIME MULTILATERALISM

Several Chinese initiatives in the Indian Ocean have stirred considerable unease among some regional powers, particularly India, which has a tendency to perceive every Chinese move in the region as a step toward its 'strategic encirclement.' Indian strategists have often argued that China's naval capability is fast growing and would soon be capable of conducting sustained operations in the Indian Ocean supported by the maritime infrastructure being built in Pakistan, Sri Lanka, Bangladesh and Myanmar (Burma).

Indian fears are accentuated by a suggestion by a Chinese admiral to Admiral Timothy J. Keating, then-chief of the U.S. Pacific Command (PACOM) of dividing the Indo-Pacific region into two areas of responsibility between the United States and China. According to the Indian press, the Chinese naval officer stated, "You, the United States, take Hawaii East and we, China, will take Hawaii West and the Indian Ocean. Then you will not need to come to the western Pacific and the Indian Ocean and we will not need to go to the Eastern Pacific.

If anything happens there, you can let us know and if something happens here, we will let you know". New Delhi has not been receptive to Chinese requests to join Indian Ocean multilateral maritime security initiatives such as the Indian Ocean Naval Symposium (IONS) and the trilateral grouping of India, Brazil and South Africa (IBSA), which has a significant maritime component in its interactions.

IONS is an initiative by 33 Indian Ocean littorals wherein their navies or the principal maritime security agencies discuss issues of maritime security, including Humanitarian Assistance and Disaster-Relief (HADR) throughout the Indian Ocean Region. The PLAN had approached the Indian Navy to 'explore ways to accommodate Beijing as either an observer or associate member'; however, New Delhi turned down the request because, in its perspective, there was 'no strategic rationale to let China be associated with IONS as it was strictly restricted to littoral states of the Indian Ocean'.

The IBSA trilateral grouping is an offshoot of the broader South-South cooperation started in 2003. Although cooperation in the security domain was not envisaged at its inception, maritime security issues (sailing regatta, trilateral naval exercises IBSAMAR, and high-level naval exchanges) have gradually gained momentum in the discussions.

China has been exploring the possibility of joining IBSA, but the fact that "IBSA's common identity is based on values such as democracy, personal freedoms and human rights" preclude its membership. In response, China craftily has attempted to dent the IBSA architecture and wean some of the actors away through bilateral political-military engagements much to the consternation of other partners.

Beijing has adopted a sophisticated strategy to build-up bilateral military relations with Brazil, and Brasilia has offered to help train Chinese naval pilots on NAe São Paulo, which is a Clemenceau class aircraft carrier. According to discussions that this author had with some Indian naval analysts, there are fears that the above collaboration could well be the springboard for reciprocity involving the training of Brazilian naval officers in nuclear submarine operations by the PLAN and joint naval exercises in the Indian Ocean. Further, these initiatives would add to China's power projection capability and could be the catalyst for frequent forays in the Indian Ocean. Although the Chinese strategy of maritime multilateralism is premised on cooperative engagements, Beijing is leveraging its naval power for strategic purposes.

The development of military maritime infrastructure in the Indian Ocean would provide China access and a basing facility for conducting sustained operations and emerge as a stakeholder in Indian Ocean security architecture. Maritime multilateralism has so far produced positive gains for China and would be the preferred strategy for conduct of its international relations in the future, particularly with the Indian Ocean littorals.

10

Arms Control Proposals And Prospects In The Indian Ocean

At the initiative of Sri Lanka, later joined by Tanzania, the United Nations General Assembly, at its twenty-sixth regular session in 1971, declared the Indian Ocean "within limits to be determined, together with the air space above and the ocean floor subjacent thereto... for all time... a zone of peace". While preserving free and unimpeded use of the zone by the vessels, whether military or not, for all nations in accordance with international law, the Declaration called on the "great powers" to halt "further escalation and expansion of their military presence in the Indian Ocean", and to eliminate from the Indian Ocean "all bases, military installations and logistical supply facilities, the disposition of nuclear weapons and weapons of mass destruction and any manifestation of great power military presence... conceived in the context of great power rivalry".

The year, at its twenty-fifth anniversary session, the General Assembly had adopted Resolution 2749, the Declaration of Principles Governing the Sea-Bed and the Ocean Floor, and the Subsoil Thereof, beyond the Limits of National Jurisdiction, which in part called for reservation of the sea-bed beyond national jurisdiction for use "exclusively for peaceful purposes" in accordance with international law, with the Charter of the United Nations, and with a new international regime to be established.

The General Assembly had, at the same session, by
Resolution 2660 (XXV), commended to member states a
treaty prohibiting the emplacement of nuclear weapons and
other weapons of mass destruction on the sea-bed and the
ocean floor and in the subsoil thereof beyond a 12-mile
zone, as defined in Part II of the 1958 Convention on the
Territorial Sea and the Contiguous Zone. The Seabed
Declaration, while including reservation for peaceful
purposes as an essential principle, was aimed at the
establishment of a resource regime, and was unlikely to
lead to detailed formulations on "peaceful purposes".

The Sea-bed Treaty was, however, in the words of one
writer, of low arms control value, in effect binding only the
superpowers and permitting "the use of the sea-bed for
facilities servicing free-swimming nuclear weapons
systems". The Peace Zone Declaration, with its roots in the
ferment of the sixties that inspired these resolutions, was
an initiative of a different order. While directed at ultimately
achieving formal international agreement on the
maintenance of the Indian Ocean as a zone of peace
(paragraph 3(c)), the Declaration was, at least in the medium
term, of an essentially political character: it was designed
to compel political focus on a region with shared
apprehensions regarding its traditional interest to the great
powers, and a sense of vulnerability in the context of the
latter's global schemes for maintaining a balance of military
capability.

It would serve as a rallying point, as a regular call to
action in the years ahead, as the states of the region grappled
with the issues involved in translating the peace zone
concept into regulatory norms and rules capable of being
administered at the national and international level.
Adopted by the General Assembly by 61 votes in favour
with none against, but with some 55 abstentions (including
all of the permanent members of the Security Council with
the exception of China), the Declaration addresses itself to
three categories of states:

(1) the "great powers", a term that must surely subsume the "permanent members of the Security Council", which are, nevertheless, mentioned separately;

(2) the "major maritime users of the Indian Ocean", or those states whose ships or goods frequently traverse the area; and

(3) the "littoral" (perhaps more generally referred to as "coastal") states and the "hinterland" states of the Indian Ocean.

The fact that there is a substantial overlap in categories - for example France, in the category of "great power", also claims through its Indian Ocean territories, to be considered a "littoral State", and some littoral states may well be categorized as "major maritime users" - appears to be of little significance. The categories essentially counterpoised to one another are the "great powers", on the one hand, and the "littoral and hinterland states of the Indian Ocean", on the other.

The Declaration makes its fundamental appeal for action to the "great powers" which are required (a) to halt expansion of their "military presence" in the Indian Ocean, and (b) to remove from the area all manifestations of their military rivalry. Such manifestations include fixed elements such as military bases, installations and logistical supply facilities, as well as mobile elements, such as ships and aircraft, to the extent that they maintain a "military presence" and are not merely engaged in transit on their lawful occasions.

It is important to note that the Declaration speaks of "military presence", which implies a situation subsisting in time of peace. The Declaration's primary aim is the elimination of any warlike presence in time of peace; the elimination of a menace to the security of a region at peace - the menace of a response to perceived threats from outside the region and unrelated to its communities; a menace that places innocent bystanders at risk and, to that extent, lacks justification on moral or legal grounds; a menace that could contribute to destabilizing or aggravating existing situations in the area, with attendant economic consequences.

The Declaration makes its second, more generalized, appeal at the same time to the "great powers", to the "littoral and hinterland states" and to the "major maritime users"; it calls on all of them to enter into consultations with a view to implementing the Declaration. Implementation of the Declaration is contemplated through the elaboration of an international agreement, which must bring into balance two elements:

(1) the prohibition, addressed principally to the great powers, of the use of ships and aircraft against the littoral and hinterland states in contravention of the UN Charter, and

(2) the right of ships and aircraft, whether military or other, of all nations, to "free and unimpeded use" of the Indian Ocean and its air space, in accordance with international law.

The Declaration finds a legal basis in the right of the states of the region to take such measures of self-defence as are appropriate in an era when the speed of ships, aircraft and weapons delivery systems makes obsolete the rigid adherence to any principle that such measures may only be legitimized in the face of armed attack.

It is not open to the usual criticism of such a thesis, since establishment of the peace zone does not itself imply the use of force but, on the contrary, the recognition of an agreed status and procedures negotiated in advance and operated in a spirit of openness and cooperation, with due regard to the legitimate rights of all states in the use of seas beyond national jurisdiction, and the air space above them.

The proposal was the result of an early concentration by the non-aligned movement on the military perils of "great power rivalry"; another proposal by Iran and Pakistan claimed that the West and South Asian regions should be nuclear-weapon-free zones; and the countries of

ASEAN in 1971 called for recognition of Southeast Asia as a zone of peace, freedom and neutrality, free from any form or manner of interference by outside powers. But the Declaration of the Indian Ocean as a zone of peace, more specific in its thrust than any similar initiative, called for a firm response from countries with global strategic concerns; and that response was far from encouraging.

THE UNITED NATIONS AND THE PEACE ZONE

The Peace Zone Declaration is directed as much to the "great powers" as to the coastal and hinterland states, and since 1971 the principal forum for concerted action has been the United Nations. The non-aligned countries gave the peace zone concept general support from the outset, as will be seen from the declarations issued every four years by meetings of their heads of state or government.

While many of them, as coastal and hinterland states of the Indian Ocean, maintain close relations with one or other of the great powers, a tilt in one direction or the other does not appear to compromise a country's non-alignment, provided that it is not perceived by the majority of the membership as being overtly partisan in a military sense exclusively in the context of "great power rivalry".

The attitudes of the great powers, however, have with few exceptions, ranged between scepticism and scarcely-veiled hostility. At its twenty-seventh regular session in 1972, the General Assembly decided to establish an *Ad Hoc* Committee on the Indian Ocean consisting of 15 members. But, while the *Ad Hoc* Committee has met regularly each year, and while its membership had expanded to 47 states by 1983, the objectives of the original peace zone concept seem as remote as ever.

Since the General Assembly's thirty-third session in 1978, the *Ad Hoc* Committee's efforts have been concentrated on convening a conference of the littoral and hinterland states of the Indian Ocean, as an essential step toward defining responses to various clauses in the

Declaration, and determining the feasibility of obtaining commitments worthy of inclusion in an international agreement.

At its meeting in 1979, the *Ad Hoc* Committee reviewed developments since the adoption of the Declaration in 1971, set out seven "Principles of Agreement for the Implementation of the Declaration of the Indian Ocean as a Zone of Peace", and outlined steps to be taken to implement the Declaration. Having been designated by the General Assembly at its thirty-fourth session as the "preparatory committee for the convening of a conference", the *Ad Hoc* Committee continues its efforts toward that end.

General Assembly Resolution 38/185, adopted without vote on 20 December 1983, called for "renewal of genuinely constructive efforts through the exercise of the political will necessary for the achievement of the objectives of the Declaration", requested the *Ad Hoc* Committee to "make decisive efforts in 1984 to complete preparatory work relating to the Conference on the Indian Ocean" with a view to enabling the conference to open in Colombo, in the first half of 1985. At its first session in 1984, the *Ad Hoc* Committee authorized the Secretariat to prepare draft Rules of Procedure for the conference stipulating, however, that those rules should provide for the taking of decisions by consensus.

While agreement on decision-making by consensus may well have eased some of the basic apprehensions of the great powers, and thus enhanced prospects for holding the conference, it may well have diminished the possibility of resolving issues of substance in the foreseeable future. The March 1984 session of the *Ad Hoc* Committee also received, but did not fully consider, a draft agenda for the conference which was proposed by Sri Lanka on behalf of the non-aligned countries.

The proposed agenda called for "Consideration of principal elements of the Indian Ocean as a zone of peace", as well as the adoption of "Modalities and Programme of Action for finalizing an international agreement and...other practical measures for the maintenance of the Indian Ocean as a zone of peace".

The "Principles" covered:

(1) Limits of the Indian Ocean as a zone of peace;

(2) Halting further escalation and expansion and eliminating the military presence of the great powers in the Indian Ocean, conceived in the context of great power rivalry;

(3) Elimination of military bases and other military installations of the great powers from the Indian Ocean conceived in the context of great power rivalry;

(4) Denuclearization of the Indian Ocean in the context of the implementation of the Declaration of the Indian Ocean as a Zone of Peace;

(5) Non-use of force and peaceful settlement of disputes;

(6) Strengthening of international security through regional and other cooperation in the context of the Declaration of the Indian Ocean as a Zone of Peace, and

(7) Free and unimpeded use of the Indian Ocean zone of peace by the vessels of all nationals in accordance with the norms and principles of international law and custom.

The problems that would confront a conference convened to deal with the "principal elements of the Indian Ocean as a zone of peace", are of a formidable order.

One can only conjecture at how representatives, with diverse and firmly held positions believed to touch vital security interests, will grapple with the meaning of terms like "bases", "military" and "manifestation of great power military presence...conceived in the context of great power rivalry"; with determining the geographical limits of the Indian Ocean as a peace zone; with satellites as an aspect of "military presence" in the zone; and perhaps, with establishing a system of verification, regarded by many as an essential element of any practical disarmament measure - all with a view to reaching a consensus on the rules to govern the conduct of states in the region.

The most recent resolution, adopted by the General Assembly at its forty-first session in 1986, calls for holding the long-contemplated conference "not later than 1988", and for completion in 1987 of the organizational and substantive aspects of the necessary preparatory work, including the rules of procedure.

The tone of conclusiveness that has characterized UN resolutions on the peace zone in recent years has not been borne out in practice, and it seems unlikely that the terms of this resolution would leave observers any the more sanguine regarding prospects for actually convening the conference. If there was a lack of significant progress at the United Nations, reflected there in an embarrassed ritualization of the demilitarization effort, the actual situation in the Indian Ocean was marked not merely by stagnation, but even by an escalation of "great power rivalry".

MILITARY PRESENCE OF THE GREAT POWERS

A great power would maintain a military presence in the Indian Ocean in time of peace for one or more of the purposes that are now part of the political tradition of such countries:

(1) to assure the security of military, commercial or fishing fleets;

(2) as a measure of national defence against possible attacks against its territories and associated interests;

(3) as a strategic deterrent in relation to other powers competing for military supremacy in the region or globally;

(4) as a visual threat or show of force, by way of support for the political penetration of a foreign country, or for the maintenance of hegemony over it;

(5) for the gathering - often clandestine - of information relevant to policy decisions concerning the foregoing; and

(6) for carrying out scientific research for commercial, military or other purposes.

The purpose for which a particular military presence manifests itself at a particular time in the area may well be lost in the convoluted military-bureaucratic processes of the power concerned. The purpose may be publicized - indeed, making the purpose public may well be essential to its achievement, the purpose may remain secret, or the purpose made public may not be the true purpose of the presence.

All that the outside world may be certain of is that, in the view of the power concerned, its military presence in the area is necessary for the protection either of its own direct interests, or of the interests of friendly states which it deems essential to protect in order to safeguard its own interests.

The peace zone concept which, in essence, is a manifestation of the right of collective self-defence evolving in response to the development of modern weaponry and the complex shifting patterns of derived political tensions, here meets squarely the claim of powers outside the region to defend their own interests which appear to be not the less vital for being far from their shores. The greater the power, the wider the range of its interests and their geographical scope.

The British military presence in the Indian Ocean, by far the most significant for over a century, declined after the Second World War with the diminishing of the Empire and withdrawal from long-held bases such as those in Aden, Colombo and Singapore. A decision by the British labour government in 1968 resulted in a drastic cut-back of its presence in the area so that a naval force that had comprised some 43 ships in 1968 had diminished to 14 by 1981. The United States presence in the Indian Ocean, however, has shown a steady increase.

Under a complex arrangement with the Government of Mauritius, the United Kingdom, by an exchange of letters with the United States in 1976, felt able to grant the latter the right to establish a "naval support facility" on the island of Diego Garcia - currently some 1 300 men at the heart of the Indian Ocean. Talks between the Union of Soviet Socialist Republics and the United States under President Carter in 1977-8 which were aimed at a staged reduction of their forces in the Indian Ocean, accomplished little and were suspended in February 1978.

The increase of Soviet influence in the area through Afghanistan and Ethiopia, and the loss of a military ally following the Islamic revolution in Iran, seemed to call for a consolidation of its own position. According to some sources, the United States currently has military agreements with Pakistan, Oman, Bahrain, Egypt, Sudan, Somalia, Kenya and South Africa on the western border of the Indian Ocean, and with Australia in the east; the core of United States forces in the Indian Ocean in 1982-83 comprised detachments of its second and seventh fleets, including two aircraft carriers with some 150 combat aircraft; two squadrons of B-52 aircraft cover the Indian Ocean from a base in Darwin, and note has been taken of the presence since the middle of 1980 of Phantom F-4 aircraft at airports in Egypt and Kenya.

According to another source, the United States by 1983/ 84 had deployed in the Indian Ocean one carrier group (some six surface combatants), nine stores ships, as well as a Middle East force (Bahrain/Gulf) of one command ship and four destroyers/frigates, and a marine amphibious unit comprising four to seven amphibious ships, with a reinforced infantry battalion group including tanks, artillery and a composite air squadron with helicopters and logistics group. Various motives have been suggested for the presence of the USSR in the Indian Ocean.

Some suggest that it perceives the four main entry points to the Indian Ocean, namely, the Suez Canal, the Straits of Malacca, the Australian coast and the South African Cape as being under the influence of the western industrialized countries, intent on maintaining their supplies of oil and raw materials, and therefore feels compelled to react by maintaining a force to protect its own interests in the area. With the nearest home base at Vladivostok some 20 000 km away, the USSR has found it necessary to set up its own network of military agreements in the area. In 1967 the USSR was granted port facilities by Somalia, and in 1969 by the People's Democratic Republic of Yemen.

In 1970 the USSR was reported to have had three submarines and six surface ships in the area. The conflict which brought Bangladesh into being saw Soviet naval strength increased to some 20 ships, and by the time of the Arab-Israeli war in 1973 to 30 ships. Following reverses in its political fortunes in Egypt and Somalia, its centre of operations shifted to Ethiopia. In 1979 and 1980 the strength of its Indian Ocean fleet was augmented by the aircraft carriers Minsk and Kiev and several support vessels which were reported as comprising some 25 military ships (including three to four destroyers, two to three troop carriers and four to five submarines) and a number of scientific research vessels.

By 1983/84 the USSR had deployed in the Indian Ocean a detachment of its Pacific fleet providing an average of two to three submarines, eight surface combatants, two amphibious and 12 support ships. In addition, it is said to maintain other forces in the area: 105 000 troops and air and armoured divisions in Afghanistan, and various military personnel in Democratic Yemen (1 500), Ethiopia (2 400), Iraq (2 000), Mali (200), Mozambique (300), Syria (7 000) and the Yemen Arab Republic (500).

While anchoring and food supply facilities are provided by India at Vishakhapatnam, Bombay, Madras and the Andaman and Nicobar Islands, and by Bangladesh at Chittagong, important Soviet facilities are said to be located in Africa: at Assab and Massawa on the Red Sea in Ethiopia; at Aden and the island of Socotra, 250 km from Capo Guardafui, and thus strategically placed with respect to traffic passing through the Straits of Bab al Mandab; at Nacala, Beira and Maputo in Mozambique; at Mahé in Seychelles; and Port Louis in Mauritius. Madagascar has reportedly permitted the location of three Soviet radar stations on its west coast, and may consider granting other facilities in small islands strategically located in the Mozambique Channel.

While the interests of France in the Indian Ocean date from colonial times, its concerns today appear to be based primarily on the assessment that the safe transport of 70 percent of Europe's oil depends on the region's stability. With the granting of independence to its Indian Ocean territories, France's military presence has shifted away from Madagascar, Djibouti and the Comoros, and since 1973 has centred on the Island of Réunion. However, some seven treaties of friendship and cooperation with its former colonies permit it to maintain a strategic position with respect to the Straits of Bab al Mandeb as well as in and around the Horn of Africa.

In Djibouti alone, France is reported to have deployed two regiments of its land forces, three combat companies, four patrol boats, and two detachments of its air force comprising twelve "F-100", nine "Jaguar" and eleven "Mirage" aircraft.

From a relatively modest force of three corvettes, three patrol boats, one troop carrier, one support ship, a tanker, a helicopter carrier and escort ship in 1976, the French military presence in the Indian Ocean is said to have increased to some 20 vessels, including ships equipped with sophisticated missile systems, and some 3 000 men.

On the basis of the information available, one seems obliged to conclude that the concept of the Indian Ocean as a zone of peace has been reduced to a pious wish, or to go even further and suggest that what was conceived as a zone of peace has actually become a zone of confrontation. By 1981, nothing had been achieved in the way of practical measures toward implementing the "Peace Zone" that had been declared a decade before.

PEACEFUL COOPERATION IN THE INDIAN OCEAN

It is against this background that one should examine the proposal of the leader of the Sri Lankan delegation to the Twenty-first Session of the Asian-African Legal Consultative Committee in 1981, H.W. Jayawardene, for exploring a new approach to regional activity in the Indian Ocean. That approach is best expressed in his own words:

"The littoral and hinterland states of the Indian Ocean share a common history of colonial exploitation and today, perhaps as a result of that common history, a relatively low level of economic development. It was in order to remove the area from great power rivalries, and ensure the Indian Ocean states the peace and security needed for their economic development, that the idea of a Zone of Peace was first conceived. It is the same goal of economic development that motivates our idea for a study of the ways and means of promoting cooperation in the management of the marine resources of this area".

In essence, Sri Lanka was proposing that, parallel with the efforts being made at the United Nations to realize the demilitarization thrust of the peace zone concept, the countries of the region, whether coastal or land-locked, should begin to explore a different, and hitherto neglected aspect of that concept - the idea of a community of states working together in a spirit of cooperation and self-reliance fostered by the emerging new Law of the Sea, to establish and maintain the institutions of peace.

The deliberate organization of peaceful cooperation by means of establishing the institutions of peace in the area was considered as having a derived arms control value when seen from at least two points of view.

The first is best expressed in the words of the writer on disarmament Elizabeth Young. Writing in 1973, she said:

> "(i) The activities of the various existing and planned United Nations bodies and of an ocean regime's own organizations are bound to result in a considerable international presence in ocean space... This presence, of itself, would have an arms control effect, proportionate to its scale and the range of its activities, and at some point it will be necessary to consider how this effect can be enlarged and enhanced... Any inspectorate, research exercise, monitoring body, is part of a *de facto* international verification system. In setting them up, the arms control significance of the information they are to acquire should be kept in view and eventually concerted."

From the second point of view, organizing peaceful cooperation in the region cannot but have the effect of promoting, through social, economic, scientific and technical contacts and exchanges, the region's sense of cohesiveness and security. Such an effect would contribute to dampening, and ultimately reducing, the tensions between individual states in the region - tensions which are the cause of arms build-up and nuclear proliferation, of the draining of slender resources into military budgets and, finally, of the subversion and weakening of the development process.

The Charter of the South Asian Association for Regional Cooperation, Preamble, paragraph 2, reads:

> "(ii) Conscious that in an increasingly interdependent world, the objectives of peace,

freedom, social justice and economic prosperity is best achieved in the South Asian region by fostering mutual understanding, good neighbourly relations and meaningful cooperation among the Member States which are bound by ties of history and culture..."

...and The Indian Ocean as a Zone of Peace, Report on a Workshop organized by the International Peace Academy reads:

"(iii) Programmes of economic and social advancement are important contributions to the maintenance of peace...":

"(v) Every step taken by the governments and peoples of the region, individually and collectively, to strengthen a regime of peace, and economic and social advancement, brings the goal set out in the 1971 Declaration closer to ultimate attainment".

States outside the region could contribute to this sense of security through offering guarantees, as they have done, for example, in Latin America by the Treaty of Tlatelolco, and through participation in a future international agreement on the Indian Ocean. Such guarantees and participation would, it seems, be more likely to be forthcoming if the states outside the region were to increase their own stake in the region's fortunes by providing finance and expertise aimed at enhancing and accelerating the organization of cooperation, through jointly-established marine-related projects and institutions.

A channelling of funds into endeavours of this type could be seen as an investment in the demilitarization of the area, quite apart from its more obvious economic development aspects. But the proposal and its timing also seemed appropriate for a more immediate reason: nearly a decade of intensive negotiations at the United Nations aimed at agreement on a new Law of the Sea was drawing to a close. An unprecedented number of the provisions of the new Convention would require parties to undertake a "duty of cooperation" toward one another, often on the regional and subregional levels. The time seemed ripe to explore the practical implications of those provisions, and evaluate their potential to make a real contribution to economic development.

FIRST CONFERENCE ON ECONOMIC, SCIENTIFIC AND TECHNICAL COOPERATION IN THE INDIAN OCEAN (IOMAC-I)

As a first step, Sri Lanka asked the Asian-African Legal Consultative Committee (AALCC) at its Twenty-first session in 1981, through a study of the Indian Ocean area, to:

(1) "determine its limits for the purpose of identifying it a special area for development;"

(2) "compile a list of national, subregional, regional and international institutions with competence or expertise in marine activities that are or could be operative in the area, as a basis for initiating cooperation and exchange of information among the States concerned or their nationals;"

(3) "consider the feasibility of establishing a consultative institutional framework for promoting the peaceful uses of the Indian Ocean, including cooperation in activities such as marine scientific research, management of living and non-living marine resources, assessment and management of environmental problems; and possibly dispute settlement mechanisms;"

(4) "carry out a survey of legal and institutional developments taking place at a national, subregional, regional or global level which have a bearing on marine activities, and are of relevance to the States of the Indian Ocean".

After presentation by AALCC of a preliminary report and supplementary information at its meetings in 1983 and

1985, Sri Lanka, having consulted the states of the Indian Ocean region, proposed the convening of a conference which would have as its objectives:

(1) creating an awareness regarding the Indian Ocean, its resources and potential for the development of the states of the region, and furthering cooperation among them, as well as among them and other states active in the region, bearing in mind the new ocean regime embodied in the 1982 United Nations Convention on the Law of the Sea;

(2) providing a forum where Indian Ocean states and other interested states could review the state of the economic uses of the Indian Ocean and its resources and related activities, including those undertaken within the framework of intergovernmental organizations, and identify fields in which they could benefit from enhanced international cooperation, coordination and concerted action; and

(3) adopting a strategy for enhancing the national development of the Indian Ocean states through integration of ocean-related activities in their respective development processes, and a policy of integrated ocean management through a regular and continuing dialogue and cooperative international/regional action with particular emphasis on technical cooperation among developing countries....

The conference, conceived as the first of a series of such conferences, was to comprise (a) a consultative phase attended by officials and experts, and (b) a final phase at ministerial level, to be held as soon as possible thereafter.

REPRESENTATION

In the event, the First Conference on Economic, Scientific and Technical Cooperation in Marine Affairs in the Indian Ocean in the Context of the New Ocean Regime (to give it its formal title) or IOMAC-I, convened in Colombo, at the invitation of the Government of Sri Lanka, and comprised the following:

> (1) a preparatory meeting at official level, 4-5 June 1985;
>
> (2) a consultative meeting attended by officials and experts, 15-20 July 1985, which took place in two stages, the first attended by representatives of the Indian Ocean states only, and a second stage in which they were joined by representatives of other states active in the region, that is the major user-states of the Indian Ocean; and
>
> (3) a final phase consisting of a meeting of officials and experts, 20-23 January 1987, which prepared the "Final Document", and a meeting at ministerial level, 26-28 January 1987, at which the Final Document was adopted.

Some 65 states, comprising 46 "littoral and hinterland" states of the Indian Ocean (LHS) and 19 "major maritime user" states (MMU) were invited to attend. Of these, a total of 48 states (33, or more than two-thirds of LHS and 15, or approximately three-quarters of MMU) were represented at one or more stages of the conference. Eighteen states were represented at the preparatory meeting, 35 at the consultative meeting and 39 in the final phase.

Of those which participated to some degree, only two states (India and Seychelles) were not represented beyond the preparatory meeting, while six others (Bhutan, Ethiopia, Kuwait, Mauritius, Singapore and Zimbabwe) did not attend the final phase of the conference. Some 18 states were not represented at any stage of the conference: 13 littoral and hinterland states - Afghanistan, Bahrain, Botswana, Brunei Darussalam, Democratic Yemen, Lesotho, Oman, Qatar, Rwanda, Saudi Arabia, Swaziland, United Arab Emirates and Zambia; and four major maritime user states - Bulgaria, Liberia, Panama and the USSR.

The United Nations and several of its subsidiary organs, regional commissions and specialized agencies which had made available expertise and financial support for organizing the conference were represented, as were several other intergovernmental and non-governmental organizations active in the field of marine affairs.

India's nonparticipation was the most notable, not only because of the country's dominant technological and political position in the region, which seemed to make a "conference on the Indian Ocean" without "India" tantamount to a contradiction in terms, but also because of reported attempts to persuade prospective participants not to attend the conference, and not to give it support or encouragement.

This initiative, which does not seem to have deterred most countries from sending delegations to one or more of the earlier meetings, may, however, have resulted in a reduction in the level of representation of participating countries at the final phase. Various reasons have been adduced for India's nonparticipation - e.g.

(1) the conference was too ambitious in scope, and thus not likely to succeed;

(2) there had not been adequate consultation in preparation for the conference, in particular with India;

(3) cooperation in marine affairs was already handled within the competences of several international organizations, so that IOMAC would only duplicate those activities;

(4) cooperation should, in any event, begin within a subregion, and then move to the regional level;

(5) the great powers and "major maritime users" should not have been invited at the current early stage of attempts to develop regional cohesion as a foundation for a cooperation framework;

(6) such a conference might have the effect of diverting concentration from, and therefore further impeding, work at the United Nations aimed at implementing the Declaration of the Indian Ocean as a Zone of Peace; and, finally,

(7) India's participation in IOMAC, a conference convened by Sri Lanka, might somehow be thought to impair the Indian Government's impartiality, and hence effectiveness, in its efforts to mediate in delicate negotiations which were taking place between the Government of Sri Lanka and the leaders of some secessionist groups in that country.

The factor of "duplication of activities" may well have gained in significance because negotiations to establish a South Asian Association for Regional Cooperation (SAARC) were in an advanced stage at the time of IOMAC's first meeting.

On the other hand, IOMAC was aimed primarily at making a reality of technical and scientific cooperation in a now familiar (if limited) field, i.e. marine affairs, without prejudicing how the initiative might eventually be thought to fit within some broader framework, such as that provided by SAARC. In any event, Article II of SAARC's Charter itself emphasises that:

"Cooperation within the framework of SAARC shall not be a substitute for bilateral and multilateral cooperation but shall complement them. - Such co-operation shall not be inconsistent with international obligations".

REPRESENTATION AT IOMAC-I

INDIAN OCEAN STATES AND OTHER STATES ACTIVE IN THE REGION

Whatever the reasons for India's nonparticipation, it was the subject of several expressions of regret and appeals for reconsideration in the closing stages of IOMAC-I. There can be no doubt that, if India had participated, she would have assumed a position of leadership befitting a country of such outstanding intellectual resources and scientific and technological achievement, and that this would have been as beneficial to the outcome of the conference, as it would have been warmly welcomed by all those present.

"[Consultations at the Conference] impressed upon us the need for the fuller participation in the work of this Conference of all the states in the Indian Ocean region and we should spare no efforts to ensure that those states which are not present today, for whatever reason, join with us in our future endeavours...". Closing statement of A.C.S. Hameed, Minister of Foreign Affairs of Sri Lanka, 28 January 1987, reproduced in Annex 4 to the Final Document of the Conference.

Organization

Officials and experts attending the Preparatory Meeting which took place in Colombo, 4-5 June 1985, carried out a survey of the most urgent needs of the states of the Indian Ocean region in the field of marine affairs, and drew up a "Tentative List of Major Items" for consideration by high level officials at an anticipated consultative meeting. The Consultative Meeting, which was convened in Colombo from 15 to 20 July 1985, dealt with the subjects on the list (1) in plenary sessions under the guidance of Hiran Jayawardene who, as the principal architect of the conference, was elected president; and (2) through eight committees on each of which all of the coastal and hinterland states of the Indian Ocean were invited to be represented.

In addition to the "Tentative List", the conference had before it a working paper entitled "Basic principles for economic, scientific and technological cooperative action in the context of the New Ocean Regime", and several sectoral papers prepared by the United Nations and its specialized agencies and other international organizations. United Nations agencies, in addition to assisting in the financing of the conference and preparing a very large volume of useful background material on the resource potential of the Indian Ocean and possible approaches to cooperative management, provided personnel to augment the secretariat of the conference.

The results of the work of the committees were considered at plenary sessions in which the coastal and hinterland states were joined by the "major maritime users" of the Indian Ocean. Their report, containing a summary of possible cooperative activities in specific subject areas, then became the principal basis for discussion at the final or ministerial-level phase of the conference which convened in Colombo from 20 to 28 January 1987.

In its final phase, the conference, having reviewed, in some seven working groups between 20 to 23 January, the range of possible cooperative activities reported on by the consultative meeting, drew up its own report containing the draft of a final document. The Final Document was then considered and adopted at a series of meetings at ministerial level from 26 to 28 January, under the chairmanship of A.C.S. Hameed, Foreign Minister of Sri Lanka.

FINAL DOCUMENT

The Final Document adopted by IOMAC-I, which takes the form (if not in express terms) of a declaration, comprises the following sections:

(i) Introduction
(ii) Preamble and Framework of Cooperation
(iii)   Programme of Cooperation
(iv)   Plan of Action
(v) Implementation and Follow-up Arrangements

The introduction amounts to no more than a protocol or chronological record of work along the lines of the final act of a conference. The rest of the document may be dealt with conveniently under three broad headings:

(i) Preamble
(ii) Modalities of co-operation (Sections II and V)

  (iii) Subject matter of cooperation (sections III and IV).

PREAMBLE

The preamble refers to the resources of the Indian Ocean, and the need to adopt a strategy for enhancing the economic development of the region through proper management of those resources. As part of that strategy, the preamble foresees the establishment of a "consultative forum" which would assist states of the region to implement and further develop the strategy, in collaboration with other states active in the region and the competent international organizations, through "enhanced international cooperation, coordination and concerted action".

With a view to dispelling any doubts that might have arisen regarding the essentially complementary and supportive aims of IOMAC-I in relation to the peace zone concept, the preamble concludes by reaffirming the commitment of states participating in the conference to the actual establishment of the Indian Ocean as a zone of peace as early as possible, and under the auspices of the United Nations.

MODALITIES OF COOPERATION

The Final Document deals with the modalities of cooperation in several provisions of section II (Framework of Cooperation) and of section V (Implementation and Follow-up Arrangements). These provisions may be classified as (a) principles of administrative or legislative action, and (b) IOMAC's institutional framework.

Principles of administrative and legislative action

Section II (Framework of Cooperation) is an elaboration of the draft of "Basic Principles" of cooperative action which had been before the conference as a basis for discussion. The framework interprets, in terms of the practical measures required to be taken by states, the duty of cooperation imposed by the relevant provisions of the 1982 UN

Convention on the Law of the Sea. Thus the duty of cooperation in this regional context involves principles regarding internal preparatory measures, such as giving the marine affairs sector a measure of priority in the national development plan; the acquisition and dissemination of information on marine affairs, and promotion of marine scientific research and application of marine technologies; the establishment of marine affairs institutions as a preferred means of achieving long-term regular cooperation; and the designation and notification of national "focal points" or offices which would be invested with competence to plan and implement cooperative undertakings at the national and international level.

The duty of cooperation also requires certain measures to be implemented externally, including harmonizing and strengthening international arrangements for managing marine resources, to be achieved through a network of national institutions and "focal points"; and cooperation expressed through solidarity at the diplomatic level to ensure that development assistance donor states and organizations accord high priority to marine affairs projects, and to secure support for marine affairs initiatives at international conferences.

As a foundation for such action, section V of the document contains a joint appeal to such organizations to "lend their strongest support" to the Programme of Cooperation and Plan of Action set out in sections III and IV.

Above, note 16. The document recognizes that the obligation to cooperate is essentially an obligation to act, through inclusion of the principle that, where a state has undertaken a duty of cooperation:

"it should at the request of another State in the region, and in any event as soon as practicable, commence consultations... with a view to harmonizing, strengthening or making any other necessary adjustments in existing arrangements, or agreeing upon new arrangements for the adoption or application of appropriate standards and measures...". Final Document, para. 2.2.4.

The duty to act proprio motu in implementing an obligation to co-operate receives further emphasis in the final paragraph of the section, which invites states of the region:

"in a timely manner, to take such measures of a policy-making, legal or administrative nature as may be necessary for the purpose of effective action in accordance with these principles".

IOMAC's institutional framework

The Final Document lays emphasis on institutions, both national and international, as "preferred means of organizing and coordinating long-term regular cooperation in marine affairs". IOMAC itself, as the "consultative forum" foreseen in the preamble, is conceived as being at the centre of a network of institutions forming an essential feature of the modalities of regional cooperation. IOMAC's institutional framework is outlined in section II, and further elaborated in section V. IOMAC's terms of reference are:

"to keep under review developments concerning marine affairs, particularly within the Indian Ocean region, to facilitate implementation of the new legal regime for the seas and oceans, and provide guidelines for coordinated, joint or cooperative action at the subregional, regional and global level, as well as to endorse specific plans and programmes and hear reports on progress in implementing plans and programmes previously endorsed by the Conference".

Final Document, para. 2.2.7. cf.: "Regional institutions can and should play an important role in improving political relationships and economic strength of participating countries".

In order to fulfil its functions and achieve its aims, IOMAC is authorized to establish a standing committee constituted on a "representative basis", as well as:

"such other subsidiary organs as it deems necessary and general or special groups for the study of specific questions as well as such other mechanisms as may be necessary for the follow-up or coordination of cooperative activities".

The document subsequently calls on the Government of Sri Lanka to continue to provide the facilities and services of a secretariat. Although this list of interests is introduced by the abbreviation "i.e.", there can be little doubt that what follows is intended to be an illustrative rather than exhaustive list of ocean-related interests. The intention of the drafters in specifying a membership of ten appears to have been to ensure that the responsibilities of the Standing Committee were clearly conferred upon, and accepted by, a representative core group, and not in any way to restrict participation, since the text goes on to declare that the "Committee shall be open-ended", thus apparently inviting participation by all Indian Ocean states and states active in the region.

The Standing Committee is to have "primary responsibility for taking such action as may be necessary for policy-level guidance for the implementation of the Programme of Cooperation and Plan of Action and for furthering cooperation through the framework of IOMAC". Members will hold office for the period (generally two years) between conferences, and will meet inter-sessionally "as often as necessary for the performance of the Committee's functions". The Committee, which is authorized to determine its procedure and agenda and the venue of its meetings, held its inaugural meeting immediately following the concluding session of the Conference on 28 January 1987.

The Conference directed the Committee to make arrangements to convene the next meeting of the Conference (IOMAC-II) "preferably within two years but no later than January 1990". The institutional structure of IOMAC is intended to be simple and embryonic in the sense that it could develop in accordance with the needs and policies of the membership of the Conference. No financial provisions were thought to be necessary, since members would bear their own expenses of travel and accommodation connected with meetings (to some extent offset by host state hospitality), projects would be funded from external sources, and the Sri Lankan Government had agreed, for the time being at any rate, to undertake the costs of maintaining a small secretariat, subject to reimbursement on a voluntary basis for any special service.

SUBJECT MATTER OF COOPERATION

The specific areas in which cooperation is envisaged are set out in section III (Programme of Cooperation) and section IV (Plan of Action) of the Final Document. In both sections, subjects are dealt with under the headings which were used at the consultative meeting to define the competence of its eight committees: marine science and ocean services, marine technology and training, living resources, nonliving resources, maritime transport, communications, management (including the law of the sea, marine law and policy, surveillance, enforcement, legal aspects of cooperation including institutional arrangements and joint ventures) marine environment.

The "Programme of Cooperation" follows closely the findings of committees at the consultative stage and is therefore an assessment in broad terms of the needs of the region and a reservoir of subjects from which specific projects may be derived for inclusion in the "Plan of Action". The "Plan of Action" represents the current thinking of the participants as to the priorities of project preparation, described as "short term", "medium term" and "long term". However, it is essentially a guide.

The more specific recommendations listed according to priority in the Plan of Action, need considerable refinement and elaboration before they become projects ready for financing and implementation; and it would be open to the membership to return at any time to the Programme pf Cooperation and select from it, for early elaboration, some subject not currently included in the Plan of Action. Final Document, the concluding note to which reads, in part, as follows: "Elements of the Programme of Cooperation would be drawn on from time to time in the context of implementation of the Plan of Action for purposes of further elaboration and augmentation (*sic*) thereof under the aegis of Standing Committee and in the context of the Framework of Cooperation".

"Baseline studies" aimed at gathering and evaluating essential oceanographic and meteorological data pertaining to the region, as well as information concerning available management and technological skills, and existing national and international regulatory mechanisms, figure prominently under several subject headings. Projects for augmenting capabilities within the region in the field of oceanographic and meteorological data collection, in particular by the use of advanced remote-sensing, and in the field of hydrography, receive emphasis under the heading "Marine science and ocean services".

Emphasis on the theme of data and information collection and dissemination within the region continues under the heading "Marine technology and training". A project aimed at establishing what could become a regional "technology bank" with subregional centres, is presented in outline. The functions of such a technology bank would not only include collecting and providing up-to-date information on available technologies, but would also extend to project-evaluation, and assistance in negotiating with suppliers of technology and in obtaining the necessary finance.

Recognizing that technology is primarily a subject engaging the interest of the private sector, the project sees such a technology bank as an honest broker in achieving the transfer of marine technology to the states of the region under fair and reasonable commercial terms and conditions.

Also emphasized under this heading is a recommendation of the UN Regional Meeting of Experts in Space Technology Applications in the Indian Ocean Region to establish a Standing Group of Regional Experts of the Indian Ocean states in Space Technology Applications under the aegis of IOMAC.

Projects referred to under the heading "Living resources" include those for the upgrading of basic scientific and technical capabilities in the region, cooperation in stock assessment, and in the use of fishery research vessels, the urgent collection of data and adoption of measures aimed at preventing large-scale tuna-catching operations from adversely affecting regional artisanal and long-line fisheries, as well as the negotiation of appropriate terms for grant of access to fisheries by coastal states: (1) to other coastal states in the region, and (2) to neighbouring land-locked and geographically disadvantaged states.

Cooperation in the investigation of offshore mineral resource potential, and sharing equipment and expertise for conducting offshore bathymetric and seismic surveys are among the projects contemplated under the heading "Non-living resources". "Maritime transport" was a subject considered to be of such importance that it was dealt with in a separate paragraph in section II, Framework of Cooperation. The central recommendation on the subject in the Plan of Action is for the establishment of regional institutional arrangements in areas such as ship ownership and investment, financing, ship management, operations and manning and, in particular, the creation, with assistance from the competent international organizations, of a regional "shipping forum" at which experts from the region could discuss questions affecting shipping, ports and related maritime industries.

Under the heading "Communications" the Final Document envisages the installation of advanced submarine cable (fibre optics) systems on a cooperative basis, as well as training programmes for personnel who would operate and maintain these and other advanced systems, and suggests cooperation in communications projects recommended by the UN Regional Meeting of Experts on Space Technology Applications.

Projects in the area of management include arrangements for harmonizing the various national laws and regulations of states in the region, coordination of a regional network of marine affairs institutions, training workshops for policy-makers, the introduction into pre-university and university curricula of oceanography and other subjects of direct relevance to marine resource management, and the negotiation of appropriate arrangements between land-locked and geographically disadvantaged states, on the one hand, and neighbouring coastal states, on the other.

On the subject "Marine environment" the Final Document contemplates, *inter alia,* cooperative action: (1) to provide, on an urgent basis, assistance and training for the development and implementation of coastal zone environment management plans through regional, subregional and national workshops, and (2) to ensure stringent monitoring and control of pollution from bilge pumping, dumping and other harmful activities.

Like most documents negotiated among several states pursuing varying policies and goals, the Final Document of IOMAC-I is untidy and frequently repetitive. However, it does clearly demonstrate the desire of the participating states to commence the type of international cooperative activity so frequently enjoined by the UN Convention on the Law of the Sea, and to explore together the possible modalities and subject areas of such cooperation.

If IOMAC develops along the lines currently foreseen by the participants, such cooperation will not be confined to the states of the Indian Ocean, but will be complemented by substantial inputs in the way of finance, equipment or expertise brought about through partnership with "states active in the region", inspired by, and carried out in accordance with, the new Law of the Sea and the terms of the IOMAC Declaration. The results of IOMAC underline:

(1) the importance of institutions in the conduct of cooperative activity, as shown by the Declaration's call for the establishment of national marine affairs institutions, the designation of national "focal points" which, by forming a regional network, are to coordinate internal marine affairs activities and receive and respond to external cooperative initiatives - and, indeed, by the creation of IOMAC and its Standing Committee; and

(2) the urgent need of most states in the region for basic information concerning marine affairs, from oceanographic data to technological and managerial capabilities, as well as the means of sharing such information so as to make it available as the basis for initiating and conducting cooperative activity.

The Declaration seems to lend support to an important principle of interpretation relative to the provisions on cooperation which form an important part of the new Law of the Sea. In the words of Sri Lanka's President J.R. Jayawardene, "An undertaking to cooperate would be nothing, if we did not acknowledge that it was also an undertaking to act". Thus, cooperation is, in a word, action and the duty to cooperate is the duty to act. Accordingly, the request of a state that consultations commence with a view to implementing a duty of cooperation could not be ignored or refused, if the requested state wished to avoid the risk of being held to be in breach of its international obligations.

States embarking upon such consultations might be expected to explore and assess potential for cooperation among them across a whole range of marine affairs questions, taking into account the regional cooperative framework already in place. Such consultations could be followed by negotiation and the conclusion of subregional arrangements. A conciliatory role might be foreseen for IOMAC's Standing Committee in the event that difficulties or serious disagreements should arise in the course of such interactions.

Compare the proposals of the late R.Q. Quentin Baxter, Special Rapporteur of the International Law Commission on the topic "International liability for injurious consequences arising out of acts not prohibited by international law", on implementation of the obligation to cooperate with a view to reducing the risk of transboundary harm, or to arriving at a fair distribution of costs and benefits of a harmful activity.

Evaluated in terms of the objectives stated when convening IOMAC-I, i.e. "creating an awareness", "providing a forum" and "adopting a strategy", a modest degree of success may, it seems, be claimed by the participants. It remains a matter of high priority, however, not only to maintain momentum through the promotional and steering role of the Standing Committee, but also to widen participation in IOMAC to include all the states concerned - those of the region as well as those active in the region - so that the initiative may fulfill to the maximum its long-term developmental goals and thereby make a major contribution toward bringing peace to the peoples of a region so often torn by conflict.

Bibliography

1. ACDA, *World Military and Arms Transfers, 1967-1976*. Washington, D.C., 1977; and *1974- 1986* (1988).
2. Ahmed, Abdel Ghaffar M. and Gunnar M. Sorbo, eds. *Management of the Crisis in the Sudan: Proceeding of the Bergen Forum, 23-24 February 1989*. Bergen, Norway: University of Bergen Centre for Development Studies, 1989.
3. Anthony, John Duke. *Arab States of the Lower Gulf*. Washington, D.C.: Middle East Institute, 1975.
4. Arnold, Anthony. *Afghanistan: The Soviet Invasion in Perspective*. Stanford: Hoover Institute, 1985.
5. Axelrod, Robert and Robert O. Keohane, "Achieving Cooperation Under Anarchy: Strategies and Institutions." *World Politics*, vol. 38, no. 1 (1985).
6. Bakhash, Shaul. *The Reign of the Ayatollahs*. New York: Basic Books, 1984.
7. Barre, Abdurahman Jama. *Salient Aspects of Somalia's Foreiqn Policy: Selected Speeches*. Mogadishu: Ministry of Foreign Affairs, 1978.
8. Bedlington, Stanley S. *Malaysia and Singapore: The Building of New States*. Ithaca: Cornell University Press, 1978.
9. Binder, Leonard. "The Middle East as a Subordinate International System." *World Politics*, 10(3) (1958):408-29.
10. Ahmed, Abdel Ghaffar M. and Gunnar M. Sorbo, eds. *Management of the Crisis in the Sudan: Proceeding of the Bergen Forum, 23-24 February 1989*. Bergen, Norway: University

of Bergen Centre for Development Studies, 1989.

11. Anthony, John Duke. *Arab States of the Lower Gulf.* Washington, D.C.: Middle East Institute, 1975.

12. Arnold, Anthony. *Afghanistan: The Soviet Invasion in Perspective.* Stanford: Hoover Institute, 1985.

13. Axelrod, Robert and Robert O. Keohane, "Achieving Cooperation Under Anarchy: Strategies and Institutions." *World Politics,* vol. 38, no. 1 (1985).

14. Bakhash, Shaul. *The Reign of the Ayatollahs.* New York: Basic Books, 1984.

15. Barre, Abdurahman Jama. *Salient Aspects of Somalia's Foreiqn Policy: Selected Speeches.* Mogadishu: Ministry of Foreign Affairs, 1978.

16. Bedlington, Stanley S. *Malaysia and Singapore: The Building of New States.* Ithaca: Cornell University Press, 1978.

17. Binder, Leonard. "The Middle East as a Subordinate International System." *World Politics, 10*(3) (1958):408-29.

18. Binyon, Michael. "West Acts as Midwife at Birth of New Leadership." *Times* (London), May 14, 1991.

19. Buzan, Barry and Gowher Rizvi, eds. *South Asian Insecurity and the Great Powen.* London: Macmillan, 1986.

20. Buzan, Barry, Gowher Rizvi, Charles Jones, and Richard Little. "The Logic of Anarchy: Neorealism Reconsidered." Working paper, 1988.

21. Cady, John F. *The History of Post-War Southeast Asia.* Athens: Ohio University Press, 1958.

22. Castagno, A. A. "The Somali-Kenya Controversy: Implications for the Future." *The Journal of Modern African Studies, 2*(2) (July 1964.): 165-88.

23. Donelan, Michael. *The Reasons of States: A Study in International Political Theory.* London: Allen and Unwin, 1978.

24. Donham, Donald and Wendy James, eds. *The Southern Marches of Imperial Ethiapia: Essays in History and Social Anthrapology.* Cambridge: Cambridge University Press, 1986.

25. Drysdale, John. *The Somali Dispute.* London: Pall Mall

Press, 1964.

26. Economist Intelligence Unit. *Country Reports* for first quarter 1991. — —. *Country Report: Uganda, Ethiopia, Somalia, Djibouti*, vol. 4 (1988).

27. Fenet, Alain. "Djibouti: Mini-State on the Horn of Africa." In *Horn of Africa: From "Scramble for Africa" to East-West Conflict*, pp. 59-69. Bonn: Forschungsinstitut der Friedrich Ebert Stiftung, 1986.

28. Foreign Broadcast Information Service. *Daily Report-Middle East and Africa*, September 30, 1980; *East Asia*, January 5, 1985; *Near East and South Asia*, May 29, 1991.

29. Galbraith, Kenneth. *Ambassador's Journal*. Boston: Houghton Mifflin, 1969.

30. Gascon, Alain. "*La Perestroïka à l'Ethiopienne: Le Pari de Mengistu.*" *Politique Africaine 38* (June 1990):121-26.

31. Horn, Robert C. *Soviet Indian Relations: Issues and Influence*. New York: Praeger, 1982.

32. Hyder, Sajjad. *"Pakistan's Afghan Predicament."* The Muslim (Islamabad), February 5, 6, 8, and 10, 1984.

33. International Monetary Fund. *International Financial Statistics Yearbook—1990*. Washington, D.C.: IMF, 1990.

34. Iraqi Embassy, Washington. Transcript of remarks by Saddam Husayn to U.S. Senate delegation, April 24, 1990.

35. Jackson, Karl D., ed. *Cambodia, 1975-1978: Rendezvous with Death*. Princeton: Princeton University Press, 1989.

36. Jackson, Karl D. and Lucian Pye, eds. *Political Power and Communications in Indonesia*. Berkeley: University of California Press, 1978.

37. Kaul, Lt. Gen B. M. *The Untold Story*. Bombay: Allied Publishers, 1967.

38. Nikki Keddie. *Roots of Revolution*. New Haven: Yale University Press, 1981.

39. Keerawella, G. B. "The Janatha Vimukthi Peramuna and the 1971 Uprising." *Social Science Review* (Colombo), vol. 2 (1989).

40. Lyons, Terrence. "Post-Cold War Superpower Roles in the Horn of Africa." Paper presented at the Second

Conflict Reduction in Regional Conflicts Conference, Bologna, Italy, May 1991.

41. MacFarlane, S. Neil. "Africa's Decaying Security System and the Rise of Intervention." *International Security*, 9 (Spring 1984.):129-30. — —. *Superpower Rivalry and Third World Radicalism: The Idea of National Liberation*. London: Croom Helm, 1985.

42. Mackie, J. A. C. *Konfrontasi: The Indonesia-Malaysia Dispute, 1963-1966*. New York: Oxford University Press, 1974.

43. Nakhleh, Emile. *The Gulf Cooperation Council: Policies, Problems, and Prospects*. New York: Praeger, 1986.

44. National Democratic Institute for International Affairs report. *1990 Pakistan National Assembly Elections*. Washington, D.C., 1991.

45. Nayar, Baldav Raj. "Regional Power in a Multipolar World." In J. W. Mellor, ed., *India: A Rising Middle Power*. Boulder, Colo.: Westview, 1979.

46. Oye, Kenneth A. "Explaining Cooperation Under Anarchy." *World Politics*, vol. 38, no. 1 (October.

47. Parmelee, Jennifer. "Angry Ethiopians Go On Anti-American Rampage." *Washington Post*, May 30, 1991.

48. Patman, Robert G. *The Soviet Union in the Horn of Africa*. Cambridge: Cambridge University Press, 1990.

49. Quandt, William B. *Saudi Arabia in the 1980s: Foreiqn Policy, Security, and Oil*. Washington, D.C.: University Press of America, 1983.

50. Raina, Asoka. *Inside RAW: The Story of India's Secret Service*. New Delhi: Vikas, 1981.

51. Rais, Rasul Bux. *China and Pakistan*. Lahore: Progressive Publishers, 1977.

52. Swami, Subrahmaniam. "Pakistan Holds the Key to India's Security." *Sunday* (Calcutta), November 13, 1983.

53. Tambiah, S. J. *Sri Lanka: Ethnic Fratricide and the Dismantling of Democracy*. Chicago: University of Chicago Press, 1986.

54. Teferra, Daniel. Letter to the editor. *Washington Post*, June 12, 1991.

55. United States Department of Energy, "Energy Information Administration". *Monthly Energy Review*,

April 1987.
56. Vali, Ferenc A. *Politics of the Indian Ocean Region: The Balances of Power*. New York: Free Press, 1976.
57. Vayrynen, Raimo. "Regional Conflict Formations: An Intractable Problem of International Relations." *Journal of Peace Research*, 21(4) (1984):337-59.

Index